Celtic Craftsmanship in Bronze

Celtic Craftsmanship in Bronze

H. E. Kilbride-Jones

The Cadbury Face-plaque

ST. MARTIN'S PRESS   NEW YORK

St. Martin's Press, Inc., 175 Fifth Avenue, New York, N.Y. 10010
Printed in Great Britain
First published in the United States of America in 1980

Library of Congress Cataloging in Publication Data

Kilbride-Jones, H E
    Celtic craftsmanship in bronze.

    Includes bibliographical references and index.
    1. Bronzes, Celtic—Great Britain.   2. Bronzes—Great
Britain.   3. Bronzes, Celtic—Ireland.   4. Bronzes—
Ireland.   I.   Title.
NK7943.K54   1980   739'.512'0899160941   80-10520
ISBN 0-312-12698-0

Contents

PART FOUR: Post-Occupation Styles

For Anita, a very patient wife

Preface

Primarily this book is concerned with craftsmanship, the craftsmanship of the bronze-smiths of Britain and of Ireland, during the period from the beginning of the first century AD until the end of the seventh century. The period covered by the text takes in a number of eras: the pre-Occupation era of the first century; the Romano-British period; and the so-called Dark Ages. What happened during these seven centuries is treated here as a continuing story, which indeed it must be, since bronze fabrication itself is a continuous story; and only external events and political and religious thinking can have any effect on the forms which were fabricated. Fashion was the responsibility of the craftsman, whose manufacturing skills were often severely tested. The people concerned are thought to have been of Celtic ancestry, and for this reason the art which they practiced is termed Celtic art. The Celts are known to have been expert metalworkers, and, from the artistic point of view, they are known to have been expert at designing bold eye-catching patterns, full-bodied and vital, and conspicuous when put alongside the artistic efforts of other people. Yet the Celt, for all his competence as an artist, had no art motifs which can be said to have been endemic to his world; so he took over a number of classical patterns, and by the exercise of his imagination, he managed to produce motifs with a compelling personality, which is at once seen to be thoroughly Celtic.

A book on this subject can only be written after long, patient and persistent study of the subject matter. Books on Celtic art normally start with the beginnings: but the early period of development has been very satisfactorily dealt with by Paul Jacobsthal in his *Early Celtic Art*. Less well considered is the period covered by the present volume, a period that is sandwiched in between the early art and that which is often referred to as the 'flowering period' of the eighth century. Many papers have been written on various aspects of the art during these more perilous times, but no single comprehensive work on the subject is so far available. The present book should therefore be seen as an attempt to correlate all the information available with appropriate comments and it also serves as an attempt at closing some of the proto-historical gaps with the purpose of setting down a continuous story of vicissitudes which, in one way or another, had some effect on development during the centuries in question. The story takes us through some of the most momentous times in the history of these islands. Thus, we begin by considering a mature art form, which becomes badly affected by the ebb and flow of people's fortunes during the Occupation of Britain by the Romans, seemingly having to go underground because of the inroads made by northern barbarians at the end of the second century; to emerge again, not too energetically, both before and after the rescript of Honorius, later to be miraculously preserved for our appreciation due to the acquisitive nature of the incoming pagan Saxon hordes. The story is both historical and proto-historical, and in it we see the story of Britain and Ireland. Had society of the times not been magnificently barbaric, lacking all sense of cohesion, it is doubtful if most of the objects considered would ever have been made at all. Vainglorious ladies titivated themselves before expensive and highly decorated refulgent surfaces, whilst the lords of the day strutted up and down, their sword-belts supporting richly ornamented scabbards, whilst their persons were protected by richly ornamented shields. Yet in Cunobelin's Colchester there was no residence fit for a king, though he had his own mint. The patrimony of these kings and nobles kept the metalworkers busy, encouraging them to give of their best, until all this was changed in a day or a week by the Roman presence. Thereafter, this patrimony was replaced by a stall in the market place, and the metalworkers entered the world of commercialism. Boom conditions must have turned some into capitalists, since it is clear that a few factories came into existence: but when the boom was at its height towards the end of the second century, warring tribes from the north brought disaster to the workshops. Only in Ireland did manufacture continue; but Irish society collapsed after the Roman Withdrawal from Britain, since it had been boosted by loot taken during raids along the British coasts.

One is aware that the Celt was willing to prostitute his art for gain, and the reasons why this was done must be examined. There was a lowering of standards and the incorporation into the art of patterns and motifs which up till this time would have been considered alien to the Celtic mind. One wonders why he was prepared to produce poor quality goods to meet these boom conditions. Only history can provide the answers to such questions; so, for this reason, historical sources have been consulted, for the purpose of providing short synopses of events which are relative to the period covered. However, connecting metal objects with events is no easy task, especially when so many are chance finds without associations. Well stratified artefacts are the answer, particularly those bronzes found on Roman sites; sites which, throughout the duration of their use, had histories of their own, and in most cases connecting with main historical events. Without information of this kind, metal objects could have little more value than a skeleton recovered from an unmarked grave, other than for their form or the decoration which they bear, both of which can provide clues to origin and date. But historical knowledge is necessary in the wider sense, when it comes to completing a background picture of the times during which the metalworkers plied their craft and had their being. For it is to these men that the story really belongs, a story of triumphs and vicissitudes, though the art itself shows few signs, other than the dropping of motifs or the incorporation of alien patterns, that the events of those days were no more than incidental.

As written down here, the continuing story may be found silently to ignore some accepted views of date and places of manufacture: but these are minority cases — many accepted views are given in the text without query. Arguments for or against expressed views are space consuming and cannot be other than academic. It is hoped that in the following pages will be found the details of a consistent internal scheme of dating with pointers to manufacturing methods and localities that will permit an easier filling-in of the niches in the corridors of time. With progress has come a realisation that the bronzes considered here are like pieces in a giant jigsaw puzzle, with more pieces falling into place all the time. The present work tries to complete the puzzle.

Acknowledgements

Professor Eric Birley and Sir Ian Richmond asked me to make a study of these bronzes more than 40 years ago. They arranged my visits to museums in the Wall area, to enable me to draw objects housed there. This help and interest is gratefully acknowledged here; as indeed too is the help given to me by directors and curators of the national and county museums, too numerous to mention. But in particular I would like to mention the names of Dr Graham Callander, Dr Adolf Mahr and Mr Arthur Deane, all alas, now deceased, of the Edinburgh, Dublin and Belfast Museums respectively, for their enthusiastic support of this project from the start. For other help and information I am indebted to Mr Laurence Flanagan (Belfast); Mrs Leslie Webster (British Museum); Mr David Brown (Ashmolean Museum); and Miss Charlotte Tagart (Newcastle). Finally, I must express my gratitude to Professor Leslie Alcock who undertook the arduous job of reading the manuscript, and made valuable suggestions; and to Croom Helm for undertaking the publication of this work.

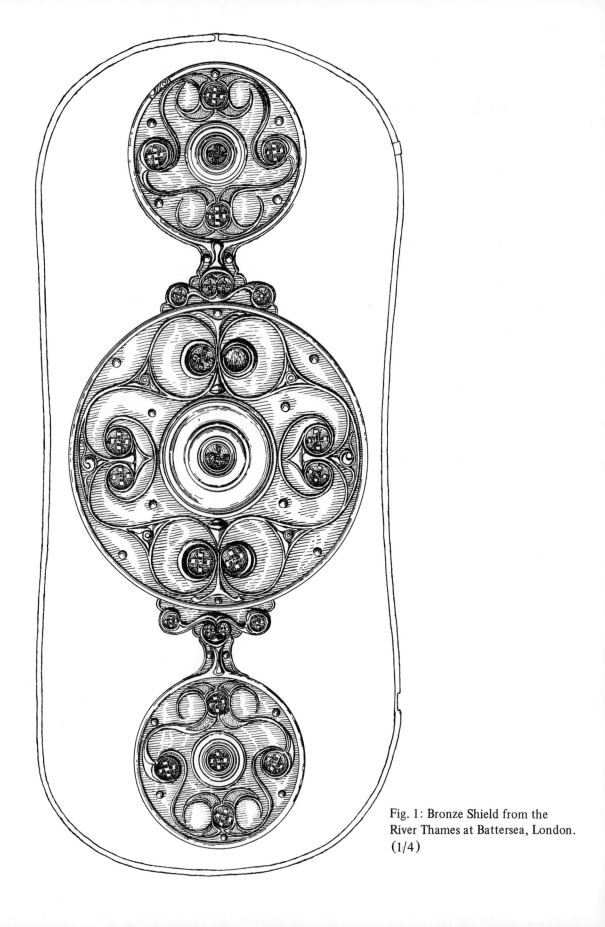

Fig. 1: Bronze Shield from the
River Thames at Battersea, London.
(1/4)

Prologue

After gold, copper is the oldest metal worked by man. Its warm, rich colour has always been highly prized. But early in the history of its use, dissatisfaction was expressed at its being too soft a metal for general use; so that ways of hardening copper occupied men's minds for some time. In the end, hardening was achieved by the addition of another soft metal, tin. The alloy produced was not only harder, but also it was found to be stronger. It was also ductile, and ideal for pouring into moulds. This new metal we now call *bronze*, and it had a lower melting point than did pure copper.

The smelting of copper, its refining and its production into 'cakes' is not a difficult process, and one that can be tackled with simple equipment. The 'cakes' are the raw material from which all non-ferrous objects were made in ancient times. Some of the copper cakes produced in Britain at the beginning of the first millenium AD have been found to be 99 per cent pure copper.[1] The metal melts at 1083°C; but the addition of tin has the effect of lowering the melting point. The molten metal was contained in clay crucibles, often pyramidal in shape.[2] The crucibles were handled with the help of tongs (Fig. 2:1): when liquid, the contents were stirred with a green branch. The methane and other gases given off would help in the refinement of the metal.

Most products were cast; and for this purpose the molten metal was poured into a baked clay mould, normally in two halves, but keyed together. There is a funnel-like opening at the top of the mould (Fig. 2:4), into which the metal was poured. The moulds themselves were fashioned on a pattern made of wood or lead, some of which still survive. When cooled, the molten metal would have taken on the exact form of the pattern. Castings are usually rough, so there is a finishing process involved. For this purpose, files, hammers, chisels, punches and so on would be needed; and, although more primitive than modern tools, their likeness is unmistakable, and in view of the quality of the work done, these primitive tools appear to have been equally effective. For decoration, the metalworker carried with him a set of patterns: today, these exist in the form of the so-called 'trial pieces'. A large collection of these was found in Cairn H at Loughcrew, Co. Meath;[3] whilst an antler object carved with patterns was found in a metalworker's camp at Dooey, Co. Donegal.[4]

It is clear that some of these metalworkers were established whilst others were itinerant. There is also evidence of the establishment of 'factories'. Metalworking had reached a very high standard of excellence at the beginning of the first century AD. Most of the objects produced then are witnesses to a maturity of form and decoration unsurpassed at any time. And all this was achieved with the help of equipment which today would be regarded as primitive.

With few exceptions, all the objects illustrated in this book are made of bronze. Most

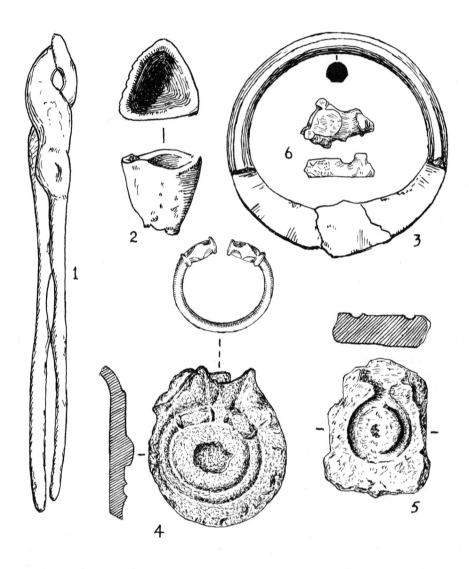

Fig. 2: 1. Tongs. 2. Crucible. 3. Unfinished Brooch. 5 Mould for Casting Bronze Ring —
all from Garranes, Co. Cork. 4. Mould for Casting Brooch, Dooey, Co. Donegal.
6. Raw Terminal Casting, Clogher, Co. Tyrone. (3/4) except 1 and 2 (3/8)

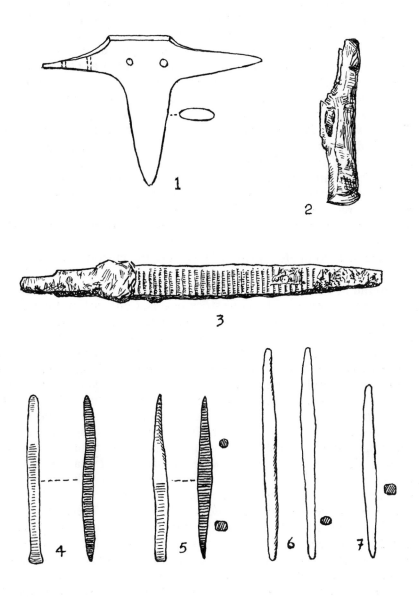

Fig. 3: 1. Double-beaked Anvil from Lusmagh, Co. Leix. 2. Hammer, Traprain Law.
3. File, Traprain Law. 4. Tracer, Lake, Wilts. 5. Chisel; 6. Punch; and 7. Awl — all from
Thames at Richmond. (3/4)

are decorated, the designs being either embossed or carried out in the linear style. Sometimes the background is plain, sometimes it is enamelled. On the other hand, some decoration has been carried out in openwork fashion. The earliest form of enamelling is that termed *cloisonné*, so called because powdered glass (enamel) was put into small compartments called cloisons, formed of upright strips of metal. Later, these cells were sunk into the metal itself, either by being cast in or chiselled out; and by the act of filling these sunken cells with glass the process became known as *champlevé*. As to the enamel itself, this is nothing more than a vitreous coating fused on to a metallic base. Enamel is closely related to glass, but it is somewhat softer, and it requires the application of less heat than does pure glass. The familiar red enamel is composed of sand, red lead and potash. First it was made into small 'cakes', and, when required, these cakes were broken up and ground down to a fine powder. A pestle and mortar were needed for this purpose: a stone mortar, in which were numerous fragments of glass, was found at Garranes, Co. Cork.[5] A third process is known as *millefiori*: basically, this millefiori enamel is made up of thin sticks composed of numerous glass fibres, the whole being fused together. The fibres are generally of different colours, and these can be arranged to form a pattern. Millefiori is normally set into a background of red enamel. Afterwards, the grinding and polishing of the surface exposes the patterns.

Metalworkers were also competent artists. They were the custodians of those art forms which, taken together, make up what is generally referred to as Celtic art. The origins of this art are to be found in continental Europe, going back to about the middle of the first millenium BC. But, at the moment when our story begins, Celtic art on the Continent had been withering for some time, the process being accelerated by the Roman advance westwards. Only in Britain and in Ireland did the art manage to continue to flourish; and such was the dexterity of the island smiths that a new vigour was given to it. But, though it survives on bronzes, obviously this could not have been the only medium for its expression: it is known, for example, that patterns were carved on wood, for some of these survive. All in all, objects made at the beginning of the first century AD are so decorated as not only to create wonder but admiration as well: gold torcs, elaborately and finely decorated with embossed designs; mirrors with their reverse sides engraved with elaborate patterns; bronze collars richly ornamented with scrolls — all these speak of affluence. Naturally, they were for the privileged, as were the decorated chariots and objects of protection and offence. The fall of Europe before the Roman advance meant that in Britain the art achieved a peculiarly characteristic insular form, developed with the benefit of British and Irish genius, and never equalled anywhere. Therefore, the beginning of our time sees the culmination of centuries of preparatory work, and it is a very good period in which to begin our story. We have a mere half century of very competent production for our delectation, before once more a Roman invasion caused a slow cessation of artistic activities. In Britain, the heroic society of pre-Conquest times was transformed into a capitalist one, and an imperial one at that. Subsequent upheavals and repressions at first had remarkably little effect on metalworking, beyond the fact that it had to be carried on in fringe areas; but, by the second century the art had to go underground for a time, which was no fault of the Romans, but of the native tribes who invaded the Province, penetrating as far south as York, and destroying the workshops en route. This happened in AD 196, but miraculously about two centuries later the art began to make a spectacular come-back.

The character of the Irish development was slightly different; and this difference is best illustrated by reference to the trumpet patterns. As was the case with Britain before the Claudian Invasion, Irish society was an heroic one, with four well-established centres interconnected by roads. Moreover, the Romans never set foot in Ireland, so that the national pastimes of feasting and fighting and looting continued as before, the four centres thriving on loot taken mostly during raids on the coasts of Roman Britain. But, on both sides of the Irish Sea, the ability of the metalworkers to improvise is well known. Whilst the Irish continued to live off Roman spoils, Britain itself had been systematically stripped of its wealth by the Romans, the population being reduced to a level a little above subsistence.[6] Yet the presence of the Roman army enabled the British to enjoy a peace they had never experienced before, particularly in the newly created Province. It is clear that, when the metalworkers had to move, they moved to Brigantia, where the pattern of life still remained unaffected by the Roman presence. But some, at any rate, were glad to make weapons for the Romans, or to make goods saleable in the forum. The result was the production of personal items, such as brooches and pins, not in hundreds but in thousands; and as the demand for these increased, so was a primitive form of mass-production introduced, which in turn led to boom conditions. The inevitable result was a lowering of standards. Yet, in the revival about the time of the Withdrawal, the metalworkers were found to be as capable as ever, indicating the continuation of skills in some remote area.

There is an aura attached to the art which is never lost on the beholder. Perhaps this is because of the skills required in the production of such minutely expressed details. It is fortunate that so many rare and excellent productions still exist, which are enough at any rate to give a clear representation of Celtic art in all its expressions and moods. Yet, seemingly diffuse, it is concocted of a small number of motifs, which remain basic and unaltered throughout its history; and these motifs can be utilised singly or in combination. However, no art can be fully understood until its basic motifs are fully comprehended. So that it is with motifs and patterns that we begin: define, and the reader gets a silhouette; but with well chosen illustrations he should get the picture in the round. If this work lives up to that intention, then Celtic craftsmanship in bronze could hope for no better method of putting its message across.

NOTES

1. J. Curle, 'Objects of Roman and Provincial Roman Origin found in Scotland', *Proc. Soc. Antiq. Scot.*, LXVI (1931-2), Fig. 37.

2. The pyramidal-type crucible was in use both in Ireland and in Britain, particularly at Traprain Law. E.g., S. P. O'Riordain, 'The Excavation of a large earthern Ring Fort at Garranes, Co. Cork', *Proc. Roy. Irish Acad.*, XLVII C (1942), Fig. 25. A.O. Curle and J.E. Cree, 'Excavations on Traprain Law in 1915', *Proc. Soc. Antiq. Scot.*, 50 (1916), p. 64, Fig. 36.

3. H.S. Crawford, 'The Engraved Bone Objects found at Loughcrew', *Journ. Roy. Soc. Antiq. Ireland*, LV (1925), p. 15.

4. E. Rynne and A.B. O'Riordain, 'Settlement in the Sandhills at Dooey, Co. Donegal', *Journ. Roy. Soc. Antiq. Ireland*, 91 (1961), p. 58.

5. O'Riordain[2], p. 138.

6. C.F.C. Hawkes, 'Britons, Romans and Saxons round Salisbury', *Arch. Journ.*, CIV (1947), p. 79.

PART ONE: BASIC INFORMATION

1. Introduction

Without their decoration, some bronze objects would seem to be rather dull. So that when we talk about craftsmanship, we think more of the artist and perhaps less of the metalworker. Yet both were in fact one and the same person. In the manufacture of Celtic-inspired products, variations in form and decoration are often considerable. This was seen as a vital antidote to boredom. Little was copied, and even upon like articles, decoration is often different. This can be put down to individualism, either the craftsman's own, or that of his client. Of course, most artists find repetitive work boring, and an arrangements of patterns is his means of self-expression. But the Celtic artist had to keep to certain rules, intended really as guide-lines: for no motif, once its form was standardised, was summarily altered. If that happened, then the motif was well on the way out.

Celtic art is abstract. Some of the abstractions became more important than did the original designs from which these abstractions were derived. For it appears that it was every craftsman's aim to see just how many abstractions he could cause to dance to the tip of his scriber. Then he stuck by them. The old sane deductions of the Greeks were set aside in a somewhat barbaric attempt at an analogy which, in the end, proved more convincing than did the original. In a lesser sense, Celtic art could be literal too: the combination of a cornucopia and two lotus buds on the same small silver brooch is sufficient enough to tell its own story. Possibly, to the Greeks, these Celtic rule-of-thumb methods may have seemed like the hermetic occultism of heretics.

Of course, many an art movement has been regarded as heretical, or as the work of heretics; but by manipulation and by the incorporation of some rationalist ideas, it has managed to become respectable. If the shield of Zeus failed to protect the classical motif from manipulation by the Celt, by that individual's very manipulation of those motifs he managed to give them an entirely new individuality. The Celt appreciated movement, so he added the tendril to the lotus-bud, the tendril itself being expressive of movement. He adapted and added, and kept on adding; and when his work was done, and was put alongside that of the classical artists, it was his which stood out as the real and rare example of technical mastery. As a result of his ability, the Celt gave to his art a distinct personality, which is impossible to confuse with another *genre*. This remarkable ability was by no means regional, but was common to all Celts throughout the Celtic world, from eastern Europe to the British Isles. Moreover, these craftsmen were intensely conscious of the aristocracy's nobility; and for objects in keeping with the noble mien, their powers of invention enabled them to produce pieces such as the Aesica brooch (Fig. 4:1). When Arthur Evans first saw this brooch he exclaimed that it was the most fantastically beautiful creation that had come down to us from antiquity.[1] Others find it vulgar.[2]

Fig. 4: *The Aesica Style* 1. The Aesica Brooch. *Embossed Disc Brooches* 2. Silchester, Berkshire.
3. Victoria Cave, Settle, Yorkshire. 4. Brough, Cumbria. 5. Aesica. 6. Brooch in Chesters Museum.
7. Richborough, Kent. (3/4) except 3 and 7 (3/2)

(1/2) except 2 (1/6)

Fig. 5: Repoussé Style Decoration. 1. Gold Torc, Broighter, Co. Derry. 2. Bronze Disc with Well, Ireland. 3. Mirror Mount, Balmaclellan, Kirkcudbright. 4. and 5. Bronze Masks, Stanwick, Yorkshire.

Beautiful or vulgar, nevertheless it symbolises the ostentation that must have been abroad in the mid first century AD.[3] This feeling of current ostentation grows with the sight of the gold torcs from Broighter, Cairnsmuir and Snettisham (Fig. 5:1; Fig. 6; Fig. 16:7). And for the lucky owners, personal appraisal was possible with the help of a bronze mirror. There were importations of Roman mirrors as well, probably less expensive than those other highly decorated props to human vanity (Fig. 7).

To the Romans, the bedecked Britons must have seemed rather more than slightly mad. 'Warlike, independent, fierce and obstinate,' wrote Tacitus;[4] 'disunion prevents their acting in concert for a public interest.' However, none of this disunion is apparent in the art, so that one must assume that the metalworkers were above political considerations. But, though a few minor representations of the human face exist, including a repoussé example from Cadbury, artistic competence did not run to statues of local heroes in the village squares. Athens and Rome were full of them, but the Britons 'had no monuments of their history'. Instead, what we have are the trappings of an heroic society, constantly reflecting a revolutionary zeal. One suspects a valedictory speech before combat; otherwise 'ruddy hair and lusty limbs' sound like a challenge to the competence of the metalworker.

The ingress of the Romans brought this heroic society to a virtual end, particularly in southern Britain. But remnants of that society continued to exist in the more remote areas of the north and west. Even before the Claudian Invasion took place, the impact of

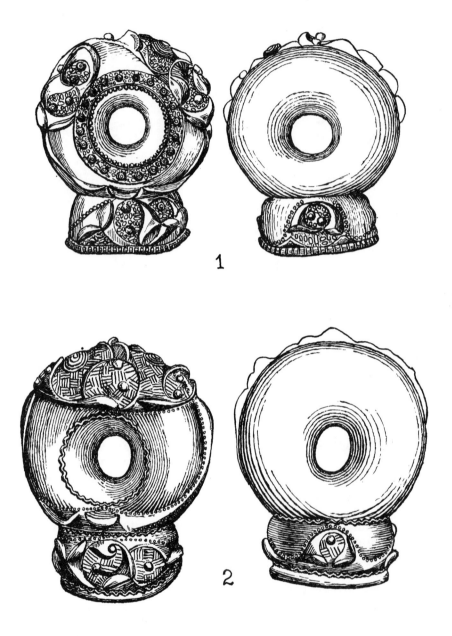

Fig. 6: Gold Torc Terminals 1. Cairnsmuir, Netherurd, Peeblesshire. 2. Snettisham, Norfolk. (1/1)

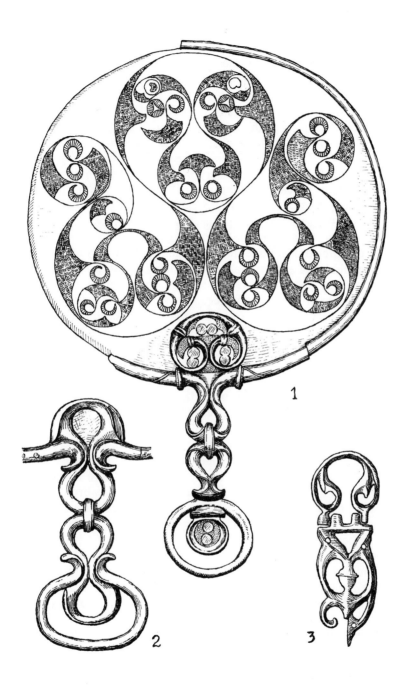

Fig. 7: 1. Bronze Decorated Mirror, Birdlip, Gloucestershire. 2. Bronze Mirror Handle, Desborough, Northamptonshire. 3. Cast Bronze Buckle, Newstead, Roxburghshire. 1 and 2 (3/8); 3 (3/4)

Rome had been felt in Britain. With the Romanising of Gaul, commercial intimacy carried the provincial Roman character to the east and south of Britain; and British nobles took the lead in acquiring Roman tastes. The Aesica brooch is a good example of an object that perhaps belonged to one of these nobles. Basically, it is a fan-tail brooch, based on an original Roman design, but it is decorated in the Celtic style. If, in some ways, Celtic art is thought of as a mystical experience, then there is nothing mystical about this brooch, or about its decoration; for in purpose it is both practical and striking. Indeed, it prepares the way for subsequent fashion. If the mating of native decoration to an alien form of brooch was seen by the craftsman as a challenge to his competence, both as a metal-worker and as an artist, then it can be said that he has succeeded magnificently. The end result could have been confusion, or at the very best hesitation: instead, there is emphasis and power in the flowing curves, suggesting an enlivened aesthetic sense. Prolonged appraisal amounts almost to a numinous experience, a final appreciation by the senses before the shadows of vulgarisation crept across the country. Thus, there was still a splendid artistic tradition at this time, happy in its vigour.

Some of the best productions are decorated in the repoussé style. This entails punching up the design from behind the thin metal sheet. Such productions are invariably spectacular, like the Torrs pony-cap, the Aesica brooch (mentioned above) and the rather unusual face-plaque shown in the frontispiece and as the jacket illustration. By making use of this technique, the Celtic craftsman found that he had better control of the medium, punching up here and there until the effect was to his liking. Thus the Cadbury[5] mask was given a lively realism, with its rather deceptive smile, which is at once pleasing. And, being a Celt himself, our craftsman stamped the object of his work with one of the current motifs, the cable motif, here used for the purpose of simulating curled hair. Not for him the plain curls of Roman fashion, even though the inspiration for the mask was Roman. It is a thorough-going piece of work, and it dates to the first half of the first century AD.[6]

The repoussé technique had long been known in Europe, but almost certainly it did not originate there: it is more likely to have been acquired from the eastern Mediterannean area before the seventh century BC.[7] By way of contrast, there is the linear style. The finest linear decoration is to be found on the reverse sides of the bronze mirrors. Abstractions are more common in linear decoration, perhaps because the repoussé style is less flexible, since a scriber can wander at will, whereas a punch cannot: in repoussé, therefore, designs tend to keep more faithfully to the older concepts. It is not clear when the first repoussé style decoration reached Britain, but it occurs on the Aylesford[8] bucket, and therefore it could be a Belgic introduction. Decorative roundels on the bucket presage those on the Broighter torc, which is of first-century manufacture.[9] The energetic sweeping curves on this torc (Fig. 5:1), or indeed on almost any object upon which they occur, are compounded of vigour allied to a wild ferocity. The basketry-filled designs on the mirrors, and linear style patterns generally, have less force: there are curves everywhere, but they are effete. The mirror-style decoration was not confined to mirrors: it appears on other articles, such as sword scabbards, spearheads and bronze mounts. This fact suggests that motifs and designs in any specific form are not of necessity linked with any one class of object.

A third way of decorating objects was in the mould. This method is still in common use today. Decoration produced in this way must be bold and flowing, though hand trimming is possible after casting. Typical of pieces produced by this method are the mirror handles. They make up into a class of their own. In their simplest forms, they are trumpet inspired.[10] But some mirrors, like those from Birdlip[11] and Desborough,[12] have more elaborate handles; the first was influenced by the trumpet motif, whilst the second was influenced by the lotus-bud motif. Both motifs were current at the same time. In the case of the lotus-buds, perhaps a certain literal connotation was intended: encouragement to stand and stare, having the wherewithal with which to do it, may have been the expressed wish of the craftsman; or else it was his little joke. However, vainglorious self-exposure before a refulgent surface is known to instil velleity, and partly this may have been the intention. The message becomes louder and clearer in the case of the little silver brooch which was found at Housesteads.[13] It is fashioned in the form of two lotus-buds springing from the mouth of a cornucopia. Nothing could be more suited to the lotus-eater. These occurrences are but minor reflections of the ostentation that was about at this time, as reflected in the Aesica brooch, the Broighter gold torc, and others similar from Cairnsmuir and Snettisham; and even the simple buckle[14] (Fig. 7:3) was in keeping with the prevailing fashion. Times were good and easy, indeed.

Such liberal use of gold, not only for adornment but as a foundation for so much artistic effort during the first half of the first century, is a measure of the wealth of the period. From the numbers of Celtic field systems existing and of storage pits, numerous south of the Jurassic Ridge[15] (Fig. 8), much of this wealth must have been measured in bushels of grain. Rich farmers must have been the economic mainstay of the country, and they were the first to suffer at the hands of the Romans. They are known to have ranged far for their requirements: in Neolithic times they bought Irish stone axes,[16] as well as others from Cumbria and Cornwall. Therefore it can come as no surprise to learn that at least one authority regards the Snettisham gold torcs as being of Irish manufacture.[17] There is, of course, a similarity of decoration on all of the gold torcs of this period. Decoration is in the repoussé style: but since decorative motifs are known to have travelled far, and that there were in existence noticeable style periods, any arguments based on decoration alone may lack conviction. More convincing is method allied to technique. All the same, Irish-made gold objects of previous ages have been known to travel far, since Ireland was a known source of gold, with an abundance of ore long since thoroughly worked out. The north-east of Scotland had been the recipient of Irish-made products from Neolithic times onwards, and this drift of Irish material things continued into the second century. Certain elements in the pattern on the Broighter torc, noticeably the cable motifs, persisted well into the second century AD, but it was their rivals, the elongated trumpet motifs, which manifested themselves on articles found in Caledonia, such as the massive armlets and the boar's head from Deskford, Banff[18] (Fig. 9:4). The basic theme is curvilinear, but the arrangement reveals an intention of apparent disorder, as with the concourse of trumpets surrounding the eye on the boar's head. There is thus good reason for believing that the north-east of Scotland received a certain amount of inspiration from the work of Irish metalworkers; and as evidence of contact one can point to the massive armlet (Fig. 9:1) which was found at Newry, Co. Down.

In Ireland, Celtic art was often less abstract than it was in Britain. It was inclined to be more decorative, with less dependence on motifs. As examples, one can suggest the sword scabbard decoration, with its basketry-filled patterns, from Lisnacrogher (Fig. 10:1, and Fig. 36). These are more decorative than are most of the patterns on the reverse sides of the mirrors, yet there is a similarity of style. The reversed symmetry of the Lisnacrogher patterns is more likely to have been continental inspired, the impetus coming from north-west France, with a stop-off at St Mawgan, in Cornwall. The St Mawgan shield mount (Fig. 10:2) was certainly under strong continental influence.[19] There is also a clear link between some of the decoration on the Stamford Hill, Plymouth, mirror (Fig. 30:1) and the running pattern on one of the Lisnacrogher scabbards (Fig. 36:1). Cyril Fox[20] placed the St Mawgan style at the beginning of the first century AD. Its secondary role in the development of British and Irish art is therefore clear. There is a measure of agreement that Irish art was always a step behind that in Britain, since the Irish received their inspiration from without. There was a large measure of borrowing from Britain; but some motifs are clearly continental.

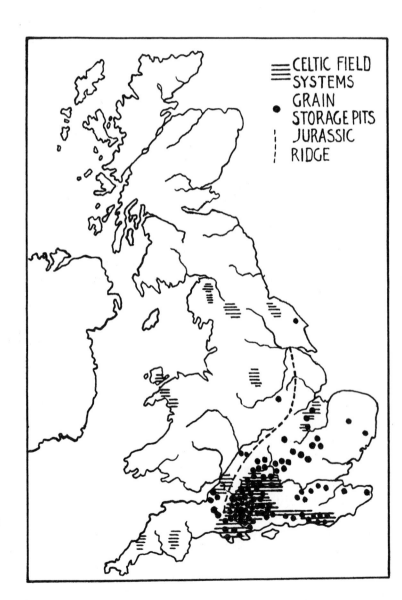

Fig. 8: The Wealth of Britain: Distribution of Celtic Field Systems and Grain Storage Pits.

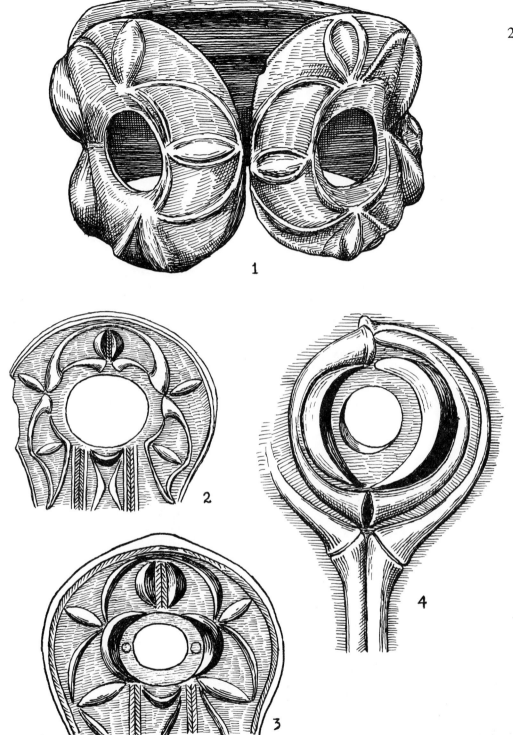

Fig. 9: Massive Armlets 1. Newry, Co. Down. 2. Detail, Belhelvie, Aberdeenshire. 3. Detail, Auchenbadie, Alvah, Banff. 4. Boar's Head, Deskford, Banff Eye Decoration. (3/4)

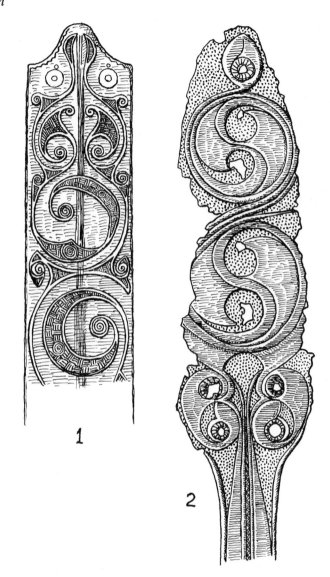

Fig. 10: 1. Sword Scabbard, Lisnacrogher, Co. Antrim. 2. Shield Mount, St Mawgan-in-Pyder, N. Cornwall. 1 approx (5/8); 2 (3/4)

The cream of early-first-century artistic effort has now been sampled, with the purpose of creating in the mind a positive picture of a Britain that was in the hands of a warrior aristocracy, well fed, well decked out, generally prosperous but completely disunited. The times were characterised by splendid works of art, every one an original — shields, scabbards, horse-trappings, chariot mounts, mirrors, torcs, collars, brooches — all adding up to wealth in craftsmanship, and all made for the warriors and nobles of the day. These were not the products of established workshops or factories (both of which were to come later), but they were the work of individuals, perhaps itinerant bronzesmiths, who were prepared to set up shop at any convenient centre, like that of Gussage All Saints, Dorset.[21]

Of such was the pattern of life in Britain before urban development took place under Roman aegis, following upon the Claudian Invasion of AD 43. After the Invasion, the old order crumbled. The people were disarmed, and the farmers were relieved of their wealth, for one aspect of life is immutable — armies of men must be fed. Therefore, the Romans requisitioned the harvest.[22] South of the Humber, from this time forward, little remained of the heroic age in Britain except for its detritus. Only in Brigantia, with its pastoral economy, did the warrior aristocracy survive, and this society existed here until AD 74. The Brigantian chiefs and nobles must have remained patrons of the old art, for it was to Brigantia that many an emigré craftsman resorted.

In what was now to become the Civil Province no real attempt was made to pick up the pieces. Perhaps there was official discouragement of any such attempt. However, under Roman rule life was to become more socially secure, more materialistic, but with an all round lowering of standards, particularly in metalwork. But some craftsmen survived through their ability to adapt, and to see through the romantic futility of the heroic age, beyond to an economic horizon which they felt it was possible for them to attain. So it can be said that their greatest achievement was that they managed to survive. The whole experience amounts to an essay in the transient. Mavericks of the trade just refused to lie down; quite soon they were offering to the public small objects which had appeal for Roman and native alike. By doing so these British craftsmen moved irrevocably into the tough world of commercialism.

Thus, British craftsmen achieved what might have seemed to have been the impossible. They switched from the manufacture of articles of offence to articles of utility, personal items such as brooches, dress-fasteners, pins and the like. None of these gave them much scope for artistic expression, because, as vehicles for imaginative effort, they were too small. But, just because they were small, they could be turned out by the hundred, if not by the thousand. So that henceforth ingenuity in artistic creation was rarely tested. This changeover was bad for the art, which now began to nod badly. Only rarely, and at odd times, did something worthwhile manage to surface. There was no encouragement towards perfection, and even less opportunity in which to exercise the imagination, since now the emphasis was on quantity, not on quality. Yet, even with this fall in standards of production, the intrinsic value of Celtic-made articles remained high, higher than that of their Roman equivalents. The Romans must have liked them as souvenirs and keepsakes, possibly because they were different. As demand increased, the artist gave way to the artisan. Most favoured of all objects was the brooch, in a hundred different forms: Roman- and Provincial Roman-inspired shapes were easily copied by the Celtic metalworkers for commercial gain. However, the northern bronzesmiths added an original form of their own design — the dragonesque brooch.

Once it was the accepted dictum that in Britain Celtic art lasted only until the early Iron Age was brought to an assumed abrupt end by the Claudian Invasion. It is now realised that the situation was more complex. Later it became popular to speak of 'an ultimate La Tène style', whatever that might mean. It is more correct to say that the Golden Age in Celtic art was halted, frozen as it were in winter's mid-stream, but later to experience a thaw, and to be revived under gentle manipulation. In Ireland, Celtic art suffered no pressures; yet there too it experienced a decline, albeit a later one. But rare plants, in benign hands, can be made to revive, and to blossom as they never did before. There was no apparent cause for the temporary eclipse of the art except for the destruction of the workshops in Britain in AD 196, and of the heroic centres in Ireland in the fifth century. That such destruction in Britain was concerted is surely proved by the fact that the Nor'nour factory, in the Scilly Isles, was also destroyed at the end of the second century. That such a temporary eclipse was possible must have been due to the fact that the art was in the hands of a select few. This was the case in Ireland, where artistic effort was confined to a small area of the country, the central belt that lies between Galway Bay and Dundalk Bay, together with a narrow strip of country in the north east that takes in Lough Neagh and the Bann Valley. All the major works of art have been found in these areas (Fig. 37). The central belt was occupied by a people called the Soghain; whilst their cousins, the Cruithin, occupied the territory from the Boyne to the mouth of the Bann. Whilst the dust was accumulating in Britain, Celtic art survived in these areas in Ireland, later to find its way to Britain in time for the Withdrawal, where it illumines the so-called Dark Ages as a result of its application to the escutcheons and prints of the hanging-bowls. This is something of an over-simplication, since the position was a little more complex, and all these complexities will be examined below.

The story of Celtic art in Britain and in Ireland would be less dramatic without some knowledge of the political and military happenings of the times. It is enlivened by the marital disagreements of Cartimandua and her consort. It has to contend with the ravaging of hostile tribes like the Novantae and the Dumnonii, and by the Caledonians who still clung to the old order. In contrast, there was the co-operative attitude of the Votadini, whose territory spread from the Tweed to the Forth, and of the Orkney chiefs, who were left with little choice when surrounded by Agricola's fleet. Both were in a position to contain the Caledonians, a name given by the Romans to those tribes who occupied the whole of the north-east of Scotland, north of the Tay. But the Caledonians maintained contact with Ireland through an ever open back door, the route via Glen Dochart, Ardlui and the Mull of Kintyre, opened up by the makers of Group IX stone axes, and which even the Romans failed to breach. In Wales, Roman soldiers marched up and down the valleys, but they left the hills to the people, as we find at Tre'r Ceiri. The same could be said for the native farmers, cultivators and herdsmen of Cumbria, of west Yorkshire, and even of those who lived on the Downs of southern England.[23] Vast areas of the territory of the Dumnonii in Cornwall and Devon appear to have been free. With all this, together with the freedom enjoyed by the Votadini in their oppidum at Traprain Law, it must be realised that there was still room for native pursuits, which must have been a continuous source of trouble to the Romans. Eruptions and repressions appear to have had little effect on the art, and therefore perhaps on those who practiced it; possibly craftsmen continued with their work by the simple expedient of ignoring these events. The attitude of the northern tribes, who brought destruction to the workshops of these men, is less understandable. The reason may have been the mistaken one of cutting off supplies to the Romans. It took an upsurge of Celtic feeling amongst the people of the Province to right these matters, in which a revival of the old Celtic religion[24] must have played no small part.

In retrospect, Celtic art is seen to be eye-catching. As with most art movements there are styles. These styles can be regional, as in the case of the Galloway style, or they may be less regional and more national. But even when national, they are phasic. Several styles are evident in the first century: the boss-and-petal is an obvious example, but it is regional in the sense that it is essentially a northern style. All may be seen as expressions of individualism in the regions in which they occur. If styles overlap, this occurrence in no way lessens the importance of either one or the other, since both are capable of existing side by side. Such individualism has resulted in objects upon which such decoration appears, being grouped together under styles, seeing that the style more or less guarantees the contemporaneity of the objects. Most of these styles are not long-lived, perhaps no more than the life-span of the man who made them. This form of study may not be new, for it is one followed by some art historians, and, as a system, it appears to work well in the present instance.

Patterns gain meaning through the use of well recognised motifs. For this reason, motifs must first be studied individually, even though their full meaning is now irretrievably lost to us. However, it is the end product which really matters; for, when each and all are assembled together, they provide a picture of an art as it is known to have existed in Britain during one of the most turbulent periods of its history. Any contribution which we can make must be one of admiring appreciation of the skills that went into the production of this art, which even today must please as much as it did at the time it was current.

NOTES

1. A.J. Evans, 'On two Fibulae of Celtic Fabric from Aesica', *Arch.*, LV (1896), p. 186.

2. J.M. de Navarro, in *The Heritage of Early Britain* (G. Bell & Sons, 1952), p. 81.

3. C.F.C. Hawkes argues for a later date in the same century, *Antiq. Journ.*, XX (1940), p. 352.

4. Tacitus, *Agricola*, XII.

5. Leslie Alcock, *By South Cadbury, is that Camelot* (Thames & Hudson, 1972), p. 167, pl. XII.

6. There is a *terminus ad quem* date of AD 61, the date of the massacre at Cadbury, for this face-plaque. The metal surrounding the face has been punched up into numerous small bosses (they are too big for dots) comparable with others, for example one found at Newstead (Macgregor, *Early Celtic Art in North Britain*, 2, p. 339) in a Flavian context. The repoussé style is an early first-century AD style in Britain, a view that is supported by the Cadbury evidence. Therefore, this face-plaque belongs to the first half of the first century AD, and it is most likely to have been an import from the Continent.

7. J. Boardman, 'A Southern View of Situla Art' in J. Boardman (ed.), *The European Community in Later Prehistory* (Routledge, 1971), p. 130.

8. A.J. Evans, 'On a late Celtic Urnfield . . .', *Arch.*, LII (1890), p. 317, Fig. 11.

9. Mlle Henry has suggested this period, with which I am in full agreement. F. Henry, *Early Christian Irish Art* (Cultural Relations Committee of Ireland, 1954), p. 17.

10. The handle of the Mayer Mirror, in the Mersey County Museum, is a fine example of this form. See S. Piggott, *Early Celtic Art*, a catalogue of an Exhibition sponsored by the Edinburgh Festival Society (1970), pp. 28, 145.

11. Ibid., pp. 28, 146a.

12. R.A. Smith, 'On a late Celtic Mirror found at Desborough, Northamptonshire, and other Mirrors of the Period', *Arch.*, LXI (1909), pp. 329–46.

13. E. Birley and J. Charlton, 'Third Report on Excavations at Housesteads', *Arch. Ael.*[4], XI (1934), p. 197, pl. XXIXe:Bl. See also Ibid., XXXIX (1961), p. 36, pl. VIII:5.

14. E.g., the Newstead Buckle. J. Curle, *A Roman Frontier Post and its People* (Maclehose, 1911), pl. LXXVI:2.

15. S. Piggott, 'Native Economics and the Roman Occupation of North Britain' in I.A. Richmond (ed.), *Roman and Native in North Britain* (1958), p. 11.

16. Large quantities of porcellanite axes were exported to all parts of Britain from factories at Tievebulliagh and Rathlin, in the north-east of Ireland. The route followed by these to the north-east of Scotland was via the Mull of Kintyre, the Clyde, Ardlui, Killin, and thence onwards as far as the Black Isle.

17. de Navarro[2], p. 77.

18. S. Piggott and G.E. Daniel, *A Picture Book of Ancient British Art* (Cambridge, 1951), pl. 61.

19. The metalworkers responsible for it could have been the descendants of refugees from Caesar's Conquest of the Armorican Peninsula in 56 BC.

20. C. Fox, 'Report on the (St Mawgan) Bronze Strip', *Arch. Journ.*, CXIII (1956), p. 33.

21. *Antiquity*, XLVII (1973), p. 109.

22. Hawkes, 'Britons, Romans and Saxons round Salisbury', p. 79. By the second century AD pit-storage had been given up altogether.

23. K.A. Steer, 'The Severan Reorganisation' in Richmond[15], p. 110.

24. As witness the building of the great temple at Lydney, Gloucestershire, and dedicated to the Irish god Nodens. Others, of a small nature, were built on disused Celtic sites, like camps.

2. Factories

Today, the word 'factory' suggests some form of mechanical productivity, as a result of which there proceeds a constant stream of objects, identical in every way, one with the other. Before the introduction of machinery, such standardisation was impossible, though repeated production by hand could achieve a credible uniformity. In addition, early aids such as moulds, fashioned on a like pattern, would give a good measure of uniformity, which would persist during the lifetime of the pattern. It is clear that skilled pattern-makers existed in Britain and in Ireland during the early centuries of our era; and, as a result of their work, some castings produced at this time can stand comparison with castings produced today by more modern methods.

Collingwood[1] was perhaps the first to hint at the existence of factories, which he did in order to account for the massive build up of brooches in second-century Britain. The presence of the Roman army, and of its souvenir-hunting personnel, gave a fillip to the production of small objects, such as brooches, eminently suitable as gifts to parents, wives and sweethearts. The Roman sense of values was perhaps more ephemeral than was that of the British, and Roman soldiery had money to spend. Goods with low intrinsic values were much in demand: as wealth increased, so did the demand. One might think that the quality of the goods might have improved, but this was not the case. Works of art were less in demand than were objects of utility, which, however, still carried a certain amount of decoration. The demand created the need for the repetition of simple, service-able forms, and suitable for everyday use. Thus, there was more metalwork about, but it was even less inspiring; for quantity production, forms had to be simple. But the Britons still possessed a better sense of artistic values than did their Roman counterparts, and for this reason, and perhaps for this reason alone, their products had the greater appeal. The metalworkers were quick to appreciate the situation, and of the commercial possibilities of their still attractive manufactured items, so they set about producing them in quantity. For manufacturing purposes, a metalworker and his mate were not enough.

Established workshops were already known at Traprain Law. Their exact number is not known, but there were enough of them to constitute a factory. Remote locations such as this were the established pattern. There were others in Brigantia, and in even more remote areas. The privileged position enjoyed by the Votadini should have given them access to the markets in the Civil Province, but, in fact, Traprain Law-made goods are noticeably absent from that area. Instead, the Votadini had to be content with markets in Brigantia and to a lesser extent in Wales. Even that market must have been lost to them when Brigantia ceased to be a client kingdom in AD 74, though a trickle of goods got through. But there were workshops situated in west Yorkshire, mostly amongst the troglodytic population, where various items continued in production into the second century. However,

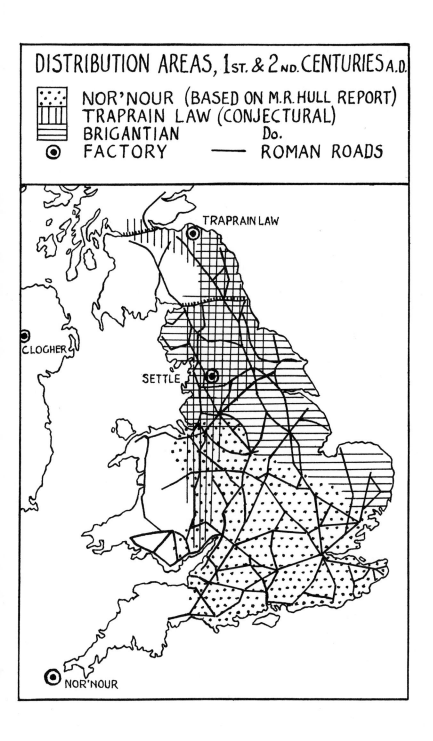

Fig. 11: Factory-made Products: Distribution Areas.

the one site which can lay claim to its having been a genuine factory site is also seen to have been the most remote of all factory sites: this is the factory formerly in production during the second century on the small islet of Nor'nour, in the Scilly Isles. It consisted chiefly of a manufacturing room and a finishing room, the latter surviving in a fair state of preservation. All of these manufacturing centres were active at about the same time, which is from the end of the first century until they were destroyed *c*. AD 196.

Nor'nour concentrated on the production of brooches, of which more than 50 varieties were made here. A very thorough study of these brooches has been made by the late M.R. Hull.[2] Two hundred and sixty brooches were found on Nor'nour itself; but Hull also studied a total number of 7,800 brooches found elsewhere in Britain, and, as a result of his studies, he was able to work out the extent of the distribution of Nor'nour brooches in the rest of the country. Hull's work, largely unsung, has provided the first positive evidence of distribution from a single factory ever recorded; and the map, Fig. 11, shows that the area covered stretched from Somerset in the south-west to Kent and Essex in the east, and then northwards as far as Leicestershire, and west as far as Denbigh, Hereford and Monmouth. A curious omission is the territory of the Dumnonii (Cornwall, Devon and parts of Somerset). The Dumnonii were Nor'nour's nearest neighbours, yet they appear to have taken nothing in the way of brooches. Presumably, the factory owners of Nor'nour would have to go to the Dumnonii for tin, and some system of barter might have been expected. However, it is clear that the Roman road system provided a means of speedy transport for the goods. Alone amongst factories, that on Nor'nour can claim to have saturated the market in the Civil Province.

From a purely business point of view, the establishment of a factory on a small islet in the Scilly Isles looks like commercial lunacy. No location could be judged to have been less suitable. With high seas running for most of the year, sea transport must have been a major problem. Not only was there the ever present problem of getting finished goods to the mainland, but raw materials for the manufacture of those goods also had to be imported by sea. Granted, tin was available near at hand in Cornwall, perhaps through the port of Falmouth; but the nearest source of copper was in north Wales, where the mines were under Roman control. On the other hand, the brooches could have been made from bronze scrap, collected during sales-drives in the Civil Province. In spite of the drawbacks, the factory managed to produce good quality goods.

The oppidum of the Votadini was also remote from the Civil Province, but it was not cut off by the seas. Also, it was not very far from a Roman road. Although the Votadini were importing goods from as far away as Gaul, they never achieved the sort of market penetration attained by the people of Nor'nour. Traprain Law oppidum housed a fairly large population, yet the volume of production never seems to have reached the figures attained on Nor'nour. This may have been for fear of over-production, for it must be remembered that the Civil Province was out of bounds to Traprain Law goods. There was a fair market in Brigantia, and a sporadic one in Wales for the same goods, so that distance was no disadvantage to fair trading. The main products were dress-fasteners and glass bangles.[3] Some dress-fasteners were sneaked into the Civil Province via the old Icknield Way, surely proof, if any were needed, that goods made at Traprain Law were banned from the Province.

A wider range of articles was manufactured in Brigantia. The workshops in the Settle area were conveniently placed. They were not very far from the Anglesey and north Wales copper mines, and there is even the possibility of a trail leading in that direction.

Amongst the wide range of articles produced are dragonesque brooches. There is good reason for believing that these were first produced in the Settle area, and for them a market penetration was achieved in an area stretching from the Forth to Norfolk. Otherwise, the main market outlets were in Brigantia itself.

An attempt has been made to show market penetration on the map of Fig. 11. Naturally, except for the known distribution of Nor'nour goods, the distribution of Traprain Law goods and of others from the Settle area is mainly conjectural. But an interesting point emerges from this study, in that, taken together, the products of all three centres penetrated the settled areas of Britain, without very much overlapping. Areas comparatively untouched are the south west of Scotland, Wales and Cornwall. Goods were probably exchanged for cash, since Roman coins are not uncommon at Traprain Law and on Nor'nour. The Nor'nour coins began with Vespasian[4] and end with Gratian.[5] There were 25 of the house of Constantine.[6] Except for a break at the end of the second century, also borne out by the pottery finds, there was continuous occupation of the islet throughout the entire Roman occupation of Britain. The late-second-century break in continuity also coincided with an abrupt end to brooch manufacture, which was later resumed on a much reduced scale.

The history of Traprain Law is not dissimilar. Regularisation of the Votadinian attitude to the Roman presence had brought a large measure of peace and prosperity to the oppidum. There was a noticeable increase in population towards the end of the first century AD. The hill was fortified, presumably with Roman acquiescence, and to this period belongs the building of the great terrace rampart, which encloses the 40-acre site. Feachem[7] thinks that this period can be equated with Fowler's second Iron Age phase, starting perhaps about 40 AD, and continuing until the last quarter of the second century. The period was one of great activity as a manufacturing town; but manufacture ceased suddenly in 196 AD, when the northern tribes burst into a province virtually stripped of troops, pillaging and destroying as they went. Thereafter, Traprain Law remained unoccupied for some 30 years or more.

Thus, the destruction of the factories at Traprain Law and on Nor'nour appear to have been simultaneous events. The story from Brigantia is similar. No new designs had been produced here after the middle of the second century.[8] The products of Brigantian workshops included dragonesque brooches, trumpet brooches, head-stud brooches as well as others, whilst cheap imitations, suitable for mass-production, appeared about AD 140 — these were a form of head-stud brooch, and they found a ready market in the Civil Province. This fact may point to another, that the quantity buyers were the Romans themselves. When the trade boom was at its height, production suddenly ceased. This event coincides with the inroads made by the northern tribes.

Thus, the sudden hiatus in production at all three centres can be blamed on the military events of the times. The Antonine Wall had been abandoned, and later the position was worsened when the governor, Clodius Albinus, transferred the bulk of the Roman garrison to the continent in AD 196, for his attempt to gain the purple. The situation was made easy for the northern tribes to burst into a virtually undefended province, and their trail of destruction has been traced as far south as York. Presumably, those who had traded with the Romans were prime targets for these people. Why the same fate was meted out to Nor'nour, and by whom, is a puzzle. Perhaps the Dumnonii were in some way involved. All this destruction constituted a major disaster for the art; and what was left of it went underground for almost 200 years, when it began to lean heavily on Christianity for its renewal.

NOTES

1. R.G. Collingwood, 'Romano-Celtic Art in Northumbria', *Arch.*, 80 (1930), p. 52.

2. M.R. Hull, 'The Nor'nour Brooches', *Arch. Journ.*, CXXIV (1967), p. 28.

3. H.E. Kilbride-Jones, 'Glass Armlets in Britain', *Proc. Soc. Antiq. Scot.*, LXXII (1937–8), p. 366.

4. AD 69–79.

5. AD 367–83.

6. AD 307–46.

7. R.W. Feachem, 'The Fortifications of Traprain Law', *Proc. Soc. Antiq. Scot.*, LXXXIX (1955–6), p. 284.

8. Collingwood[1], p. 52.

3. Forms and Motifs

The mind operates entirely through the senses. Form and decoration, incorporating motifs of standardised shapes, are therefore both concerned with the sense of sight. There is nothing new in both being manipulated in order to please the eye: amongst men, such has been the intention since the beginning of time. Forms are normally evolved out of current styles, whilst decoration in some cases includes patterns going back for hundreds of years.

For definitions, we look to simple terms. Thus, we can say that form is nothing more than the visible aspect of an object. But form is transmutable, for the reason that nobody is ever really pleased with his own creation, let alone with those of others; and dissatisfaction of this nature leads to modifications being made. Yet, modification, in its turn, may result in the production of less satisfying forms; less satisfying, perhaps, than was the case with the original form from which all these efforts started. Commercial interests can also have had a hand in these modifications, perhaps in the hope of creating new markets or reviving old ones. Continuous modification makes up into a very satisfying typological sequence, indicating how the vulgarisation of the original form can come about. So, objects can become too large, or too small, or too ornate, or not ornate enough; or the form gets so much out of hand that in the end everybody tires of it, and it is scrapped. At this point, a new form is required, and the whole process starts over again.

Decoration is normally treated differently. For it to have meaning, it must incorporate at least one motif. Unlike form, motifs are hardly ever changed: this is because motifs have meaning, and change could destroy that meaning by making the motifs unrecognisable. In Celtic art, most motifs were borrowed from classical sources, and then modified by the Celts into a form which they could appreciate, and were then standardised, and treated with respect. The Celt was a clever adaptor, but weak as an innovator. Throughout the Celtic world, a small repository of motifs was created, and this was drawn upon according to requirements; and respect was accorded to these motifs because they appear to have been talismanic. Doubtless, there was a message somewhere in each of them, but all knowledge of this has been lost to us. In Britain respect for motifs waned with the advent of the Occupation. Finally, in Christian art they had no place at all.

Multiple variations of form can bear identical decoration, though such occurrences are rare. The development of one does not of necessity have to be in step with the other, since form and decoration do not travel together. The two can come from entirely different sources. Forms evolved can be and are manipulated to suit the type of decoration applied to the piece; for they are more individual, and therefore are more ready to be moulded by the metalworker who first thought of them. On the other hand, decoration has often been worked out by somebody else, for the simple reason that it is more often borrowed than contrived by the man using it. The lives of some motifs were excessively long, merely because everybody knew of them and wanted them on his metalwork. Similar remarks apply to Christian symbols, or to the symbols of other religions, where conservatism rules the day. So it will be seen that the life of a motif could cover the life-spans of many forms, without undergoing any changes itself. The two together, in the same object, must be seen as an expression of skill on the part of the craftsman. His purpose in producing the object was that of pleasing his customer, who may have asked for the inclusion of one motif or another. Motifs, therefore, are the elements within the pattern, which the craftsman was free to alter, so long as the motifs remained untouched. Held together and in consort they can be made into almost any pattern. So that, *in toto*, motifs and patterns can be said to make up the whole complex of Celtic art.

Therefore, patterns must be broken down into their main constituents, which generally consist of one or more of the following:

(a) The lotus-bud motif
(b) The palmette
(c) The continental trumpet motif
(d) The cornucopia motif
(e) The cable motif
(f) The spherical triangle
(g) The triskele.

Palmettes, trumpets and triskeles have been the subject of much speculation in the past; and Paul Jacobsthal[1] has charged archaeologists with a complete lack of understanding and confusion with regard to the recognition of these motifs. It is not that archaeologists are insensitive to their impact, but lack of understanding can be attributed to an ignorance of their meaning, which is now lost. Mostly, the lotus-bud has been confused with the palmette, whereas both are different from one another, though both were derived from classical sources. Similarly, the trumpet has been confused with the cornucopia. Of all the motifs named, the meanings of only two are known: the cornucopia is, of course, the horn of plenty, normally represented in modern graphic style as overflowing with flowers, fruit and corn. In the Housesteads brooch it is represented as overflowing with lotus-buds, so that even here there are variations of interpretation. In ancient Greek legend, the lotus is represented as inducing luxurious dreaminess and a distaste for the active life. It appears on temple friezes, as does the palmette. The motif of most common occurrence in Britain and Ireland is the cable motif, but again nothing is known of its meaning. The same can be said of the rest of those motifs which are listed here.

(a) The Lotus-bud Motif

This motif was derived directly from Greek art (Fig. 12:2). Stylistically, the classical representation does not exactly resemble the *nymphaea lotus*, suggesting an abstraction. It was this abstract form which attracted the attention of the Celts, but its adoption was concerned only with the petals and not with the stamens. This is understandable: the stamens were somewhat geometric in appearance, whereas the petals were curvilinear. The original design had been painted into a frieze on the old Parthenon at Athens.[2] Seen through Celtic eyes, it looked different, and it became incorporated into the Waldalgesheim style, represented on the large gold torc.[3] By its incorporation into this style, the

Fig. 12: Lotus-Bud Motif 1. Decorated Bronze Plate, Comacchio, Italy. 2. Detail from Frieze, Old Parthenon, Athens. 3. Detail from Crescent-shaped Mount, Brunn-am-Steinfeld, Austria. 4. Design on Scabbard Top, Verulamium, St Albans, Herts. 5. and 7. Two Halves of Bronze Buckle, Newstead, Roxburghshire. 6. Detail from Decoration on Aesica Brooch, Northumberland. 8. Detail, Silver Brooch, Housesteads, Northumberland. 9. Detail from Belt Mount, Castlelaw, Midlothian. 3 − 9 (1/1)

respectability of the motif was assured. To give movement to it, a tendril was added to each petal. Whereas the classical method was to represent the petals in opposition, the Celts decided on linking them at the top[4] (Fig. 12:3).

Once an alien motif had been adopted into his art style, the Celtic craftsman then decided to indulge in an intellectual pleasure of intensive refinement of what he regarded as logical pattern making.[5] On the Continent, that stage was reached after the inclusion of the tendril, which, by its very nature, is representative of movement. Later, this movement was translated into circles, spirals and flowing curves, all representative of the tendril's ability to twist and turn, and to come back again upon itself. In well thought out designs, these curves provide the linkages, and in this there is some similarity to the Greek idea: stem to stem at Waldalgesheim and elsewhere, in what amounts to well set out if somewhat conservative pattern-making. But, whereas in Greek art these link lines are of a somewhat geometric nature, by contrast the Celtic artist, fancy-free, has allowed his mind to stray into concocting a comma-leaf-expanding-into-a-fanshape-at-the-pointed-end motif, which is shown in Fig. 12:1a.[6] Jacobsthal[7] has demonstrated that this fan-shaped expansion is nothing more than a closed-up form of the classical sprung palmette (compare Fig. 12:1a with 1b). To be more precise, it all harks back to the conception of the newly sprung lotus-bud painted into the frieze on the old Parthenon.

This bronze plate from Commachio was the work of a transalpine bronzesmith,[8] and it would appear to be of early date. Though occurring singly here, elsewhere this comma-leaf-expanding-into-a-fan-shape-at-the-pointed-end motif became the basis for linked patterns as indicated above; but now the curved linkages, instead of being of even thickness, as in the Greek examples, are seen to swell out in typical comma manner.

The Commachio plate has been discussed here because it was at the beginning of a fashion in linked patterns. Britain was the recipient of this art style at a much later date, at a time when the Art movement was slowing down on the Continent. The British craftsman, and with him the Irish craftsman, was less conventional, with the result that an immediate sense of freedom is noticeable — freedom to adapt, coupled with an abandonment to swinging movement, and an almost total riddance of some of the disciplines. The result is that, in Britain and Ireland, movement is more free-flowing, as may be seen in the case of the shield mount (Fig. 10:2) from St Mawgan-in-Pyder, north Cornwall.[9] In contrast to the original method of representation, here the tendrils have been stretched and curved to enclose a pair of lotus-bud motifs in opposition, and then to conjoin them to a second pair in a movement so free of restraint that it amounts almost to an abandonment of self-discipline. Fox[10] thought that this piece might have been the work of craftsmen working at the southern end of the Jurassic zone, with a manufacturing date at about the turn of the century BC–AD. The style is definitely continental and not British, though the wild freedom of movement is British and not continental. The metalworker was probably descended from someone who fled the Armorican peninsula in 56 BC, at the time of Caesar's conquest of that region.

Another mount that comes nearer in character to the continental conception of Celtic art than it does to the British is that shown in Fig. 23:1, from Polden Hill, Somerset. The Polden Hill hoard[11] contained 14 bridle-bits of excellent workmanship, some having circular sockets filled with enamel; enamelled cheek-pieces, one of which is illustrated in Fig. 17:4; an embossed shield boss; three brooches of Collingwood Group E form;[12] and champlevé enamelled horse-brooches or mounts,[13] two of which are shown in Fig. 23:1 and 2. The larger of these two mounts is ornamented with a design on an enamelled background, and among the constituent parts of this design there are representations of lotus-bud motifs: two are of the paired type, as shown on the crescent-shaped mount from Brunn-am-Steinfeld, and again at Waldalgesheim, but here separated. The other two representations of the motif are very good copies of it as it appeared on the frieze at the old Parthenon in Athens. So, what we have here is a rather remarkable occurrence of two forms of representation, one being a fair copy of the original form as seen amongst Greek art. How this came about is not clear;[14] but the British preferred the classical form.

Fig. 13: Lotus-Bud Motif Decorative Details 1. and 4. Torrs Pony-Cap, Kirkcudbright. 2. Turoe Stone, Co. Galway. 3 Trumpet Mouth, Lough-na-Shade, Co. Armagh. 5. The Cork Horns, Cork. 6. Sword Scabbard, River Bann at Toome, Co. Antrim. (Not to scale)

The lotus-bud motif is also found in the Galloway region of Scotland, where there is known to have been a first-class metalworking tradition up till the first century AD. Here we find a decorative style with some strong similarities to decoration known to have come from the territories of the Soghain and the Cruithin in Ireland. This suggested cross-channel relationship becomes more positive when analogies are found on the Turoe stone at Loughrea, Co. Galway[15] (Fig. 13:2), for the rather remarkable decoration on that equally remarkable production, the Torrs chamfrein,[16] which is really a pony's head-piece. Two decorative details have been taken from it, and these are shown in Fig. 13:1 and 4. The whole of the Torrs pattern is carried out in repoussé style, and it is quite distinct from anything else seen in other parts of Britain. The repoussé style was perhaps more widespread in Ireland than ever it was in Britain, where its occurrence was in areas under Irish influence. Representations of the lotus-bud motif are in two different forms on this pony-cap: the first is robust, forceful, whilst the second (Fig. 13:4) is more effete. Here again, the genesis of the style could be looked for on the decorated bronze plate from Commachio. A point to remember is that in Ireland, in the first century BC, there was a living and a vital art style, which can stand comparison with the best that the Continent or Britain has to offer by way of comparison; but that this Irish art was not inspired from Britain, but from the Continent. The paucity of finds has caused people to make statements which cannot be maintained, thereby creating the impression of a weak Irish involvement. But Ireland had its first sight of real Celtic art when the only genuine piece in the Waldalgesheim style ever to have been found anywhere in the British Isles — the gold torc from Clonmacnoise[17] — was first brought to Co. Offaly. Then there followed sophisticated forms of decoration which must have been widely current, as witness their occurrence on the Turoe stone.

Another high quality Irish production is the bronze trumpet from Lough-na-Shade, Co. Armagh,[18] with its elaborately decorated mouth. Part of the pattern of decoration is shown in Fig. 13:3, and from this drawing it will be seen that one finial in that decoration has been given a twirl, which is said to resemble a snail-shell. Such a representation presages its further development, and its later occurrence on the Broighter torc (Fig. 5:1). Amongst the decoration on the trumpet mouth there are curves which are fast becoming tenuous, and this tendency in turn affects representations of the lotus-bud motif. However, the relationship of the trumpet mouth lotus-bud motifs with those on the Torrs pony-cap remains clear. Such a general style is reflected weakly on the Newnham Croft, Cambridgeshire, bracelet.[19] Atkinson and Piggott[20] date this style too early. A more acceptable date would be some time in the first century BC. As a final observation on relationships, it can be stated that the Cruithin of Galloway and the Cruithin of the north-east of Ireland were one and the same people.

There was also a second style present in Ireland. This is the linear style, well represented at Lisnacrogher. Here the decoration on the sword scabbards is carried out in this style, as indeed it is on another scabbard from the River Bann at Toome[21] (Fig. 13:6). This Irish linear style is more akin to the art style of south-west Britain, whilst the basketry filling to the design on one of the Lisnacrogher scabbards is a definite link with the mirror style. On the Irish sword scabbards the lotus-bud motifs are represented singly, in the Polden Hill manner. The Waldalgesheim tradition of pattern-making began a movement that is well represented at Polden Hill, on both the larger and the smaller mounts: but, whereas the art on the Continent was already on the way out in face of Roman pressure, in Britain and in Ireland it was still aggressively alive. Already the remnants of the Waldalgesheim style were being faded out, for they had survived for too long. But many of the old motifs were to be retained, even though they underwent a change of style, as a result of revitalisation in Britain.

The Polden Hill finds belong to the first century AD. The decoration on the larger of the two mounts (Fig. 23:1) is seen to include a more classical representation of the lotus-bud motif, and this could have been its first introduction into Britain. It would seem that all the continental Celtic adaptations had been bypassed, and that a return was made to the basic form in an effort to be rid, once and for all, of the Waldalgesheim development. This may be put down to an expression of British independence. The more classical form of the motif also served as a basis for some of the mirror decoration.

After AD 43, the date of the Claudian Invasion, there was a noticeable shift of artistic effort to the north, mainly Brigantia. There follows a rash of objects, many bearing bits and pieces of recognised patterns and motifs, amongst which it is possible to recognise the lotus-bud. Perhaps the best representations are those on the Aesica brooch. Here two motifs have been incorporated into the pattern on the arch of this brooch, and it will be noted that they have been conjoined at the top (Fig. 4:1). A somewhat similar arrangement is seen on the Newstead buckle[22] (Fig. 7:3). On the lower half of the buckle there is an interesting deviation, in that the lower tips of the petals are seen to be conjoined, springing as they do from a miniature urn, a firm indication of Roman influence. This buckle was made some time during the early period of the Occupation, possibly before Newstead went up in flames shortly after AD 105. Elsewhere, the Aesica brooch has two lotus-buds in opposition (Fig. 12:6). Decoration is in the repoussé style, whereas the Newstead buckle is in cast openwork.

Well executed representations of the lotus-bud motif appear on a small silver brooch which was found during the 1898 excavations at Housesteads.[23] Here they spring from the mouth of a cornucopia (Fig. 15:7 and Fig. 12:8). Quite clearly there is a literal meaning attached to this arrangement. Another brooch, also cast, is fashioned in the form of three cornucopias and one lotus-bud: this is the one (Fig. 15:9 and Fig. 12:9) which was found in the souterrain within the fort at Castlelaw, Midlothian.[24] And at Alnwick Castle there is a cast brooch,[25] the design on which incorporates lotus-bud motifs, and in style they are not unlike those on the Aesica brooch.

(b) The Palmette

The palmette is a conventional palm-leaf design which appears in stylised form on many Greek buildings, particularly temples. It also appears as an integral part of the decoration on some Celtic metalwork: but occurrences in Britain are on a more limited scale than they were with the lotus-bud motif. Reginald Smith was of the opinion that the barbarian artist had the opportunity but not the power of copying the drooping palmette. Opportunity came with the motif's appearance on a number of Greek wine flagons and pottery types which were exported to central Europe. It is easy to see why the lotus-bud motif gained in popularity over the palmette: it has flowing, sweeping lines, with which the palmette's almost geometric form compares badly. Finally, the Celtic artists adopted the enclosed palmette, for the reason that the enclosures are curvilinear, with spiral-like ends (Fig. 14:1 and 2). Quite quickly the palm leaves were shed until there was nothing left by the curvilinear enclosures. These developments are set out in Fig. 14.

Reference to the gold disc from Auvers[26] will serve to indicate how the curving enclosure had become the artists' main consideration, and that these curves and spiral-like ends were to suggest the course which future modifications would take. However, these forms, as set down here, were only at the beginning of development: at Besançon,[27] in a band of running palmettes, the spiral ends have been eliminated. The palmette-forms seen on the bronze hanging-bowl from Cerrig-y-Drudion, Denbigh,[28] are seen to be very similar (Fig. 14:6). This is the earliest record in Britain of a continental-inspired modification of the classical palmette. It took a century for the motif to achieve this devolved form: for the Besançon wine flagon belongs to the end of the fourth century BC whilst, at its latest, the Cerrig-y-Drudion hanging-bowl is most likely to be of third century BC date. The importance of the decoration on this hanging-bowl to the history of Celtic art forms in Britain cannot be over-emphasised: for, not only is there a direct link with the Mediterranean art movement, but motifs and patterns hitherto unknown in Britain were here introduced into the country.

One wonders if the palmette, as a symbol, had anything to do with Bacchus, in view of its representation on wine flagons. If so, then the Bacchantic representations on the Besançon wine flagon take three different forms. One has already been mentioned. Another is shown in Fig. 14:4. This one is not too far removed from the classical representation. Of interest is the entwining of the lowest leaf on both sides with the expanded end of the enclosing scroll, an idea that was carried a stage further in design on the same vessel, and is also remarkably like the palmette on the bronze disc from Écury-sur-Coole, Marne[29] (Fig. 14:7). Entwined ends become like double spirals, whilst the number of leaves is reduced to one. There are fair equivalents in Britain on the Battersea shield (Fig. 14:8). Here, however, the design has been carried out in repoussé style, a style which gives to the motif a somewhat different character: for now it takes on a boldness never before achieved in linear decoration. The double spiral ends are here dispensed with, and their places are taken by coloured glass inserts relieved with swastika patterns. The Battersea shield used to be generally thought to have belonged to the turn of the century, BC–AD. However, it is quite possible that it is of later date, on the evidence of the repoussé work itself and of its style.

Final development of the motif entailed drastic simplification. What were regarded as unnecessary details were now dropped, in an effort to achieve a bold outline. The end result of this treatment is represented by a form which was ideal either for repoussé work or for casting, and this could have been the intention. This new 'decapitated' palmette (as Powell[30] has termed it) is seen to its best advantage on the bronze horn-cap (Fig. 29:2) from the Thames at Brentford.[31] This much simplified form of the motif must have been quickly developed. It could have been a Belgic abstraction, in view of its association with the handle of the Aylesford bucket.[32] A detail taken from amongst the decoration on the Brentford horn-cap is shown in Fig. 14:9. Quite clearly, the Battersea shield palmettes are not far ahead of this ultimate style.

The palmette is noticeably absent during the Romano-British period, but it reappeared after the Withdrawal. In post-Roman times it is represented in the form of voids, and chiefly on the hanging-bowl escutcheons, such as those from Baginton, Warwickshire;[33] from Tummel Bridge, Perthshire;[34] and from Castle Tioram, Moidart;[35] and they also appear on a bronze plate from Richborough.[36] The Irish made use of an even more simplified form, which is of common occurrence on the terminals of zoomorphic penannular brooches.

Fig. 14: Enclosed Palmette Details from 1. Vase from Cyprus. 2. Vase from Kamiros, Rhodes. 3. Gold Disc, Auvers, Seine-et-Oise. 4. and 5. Bronze Wine Flagon, Besançon. 6. Decoration on Hanging-Bowl, Cerrig-y-Drudion, Denbigh. 7. Bronze Disc, Écury-sur-Coole, Marne. 8. The Battersea Shield. 9. Bronze Horn-cap, Thames at Brentford. (Not to scale)

(c) The Continental Trumpet Motif

Trumpets have never been correctly differentiated. One form has a long history, for it was first noted on the crescent-shaped mount from Brunn-am-Steinfeld (Fig. 12:3). But here it is still in embryo form. The same form can be seen also on the Waldalgesheim torcs. The motif is easily distinguishable, because its thinner end is always partly (not

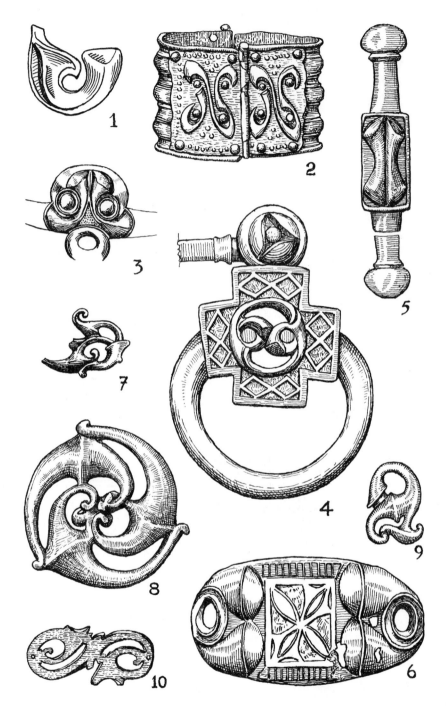

Fig. 15: Continental Trumpet and Cornucopia Motifs 1. Detail from Aesica Brooch Decoration. 2. Bronze Armlet, Plunton Castle, Borgue, Kirkcudbright. 3. Terminal Loop of Handle of Bronze Mirror, Holcombe, Uplyme, Devon. 4. Bronze Enamelled Bridle-Bit, Thames at London. 5. Bronze Cheek-Piece, Birrens, Dumfries. 6. Bronze Belt-Plate, Drumashie, Dores, Inverness. 7. Silver Brooch, Housesteads, Northumberland. 8. Bronze Mount, Icklingham, Suffolk. 9. Bronze Belt Mount, Castlelaw, Midlothian. 10. Bronze Belt Mount, Corstopitum, Northumberland. (3/4)

wholly) curved round a small boss. For this reason it is often confused with the cable motif. This trumpet motif had already had about 400 years of history behind it when, for the first time in Britain, it appeared on a bronze armlet found at Plunton Castle, Borgue, Kirkcudbright.[37] Here it makes up one half of a design which Leeds has termed a 'swash N'. On the armlet there are two panels with decoration (Fig. 15:2), this decoration having been punched up from behind by means of a die. This must be one of the first instances of competent die-work known in Britain. Should the trumpet motifs here be compared with those on the Brunn-am-Steinfeld mount, it will be noted that apparent differences are extremely small. On the armlet, the trumpet mouth tends to be triangular in cross-section, but straight cut across the mouth. These facts are important, since they are repeated on the Aesica brooch (Fig. 15:1), in which case the triangular form is quite deliberate. On the Aesica brooch the motif is coupled to a lotus-bud. Motif combinations of this nature are not uncommon in Britain, though rare elsewhere.

The Plunton Castle armlet is indubitably an import of the first century BC, and it was found in Galloway, a recognised metalworking centre, with strong affiliations with the industry in Ireland. But the Irish form of trumpet differed from anything seen in Britain, in that it is elongated, at times into sinewy curves. Moreover, the mouths of the trumpets are 'capped'. The Plunton Castle and Aesica trumpets are short, and sometimes they are foreshortened as they appear on the Holcombe mirror handle (Fig. 15:3), which comes from Uplyme, Devon.[38] But at Holcombe, though the trumpets remain of triangular cross-sectional shape, the mouths are domed; and another development is the conjoining of trumpets at their thin ends. Even so, partly they still curve round a central boss, so that at least one old tradition is carried forward. Twin trumpets form minor roundels within a larger one on the bridle-bit from the Thames at London[39] (Fig. 15:4), but now the boss has been dropped, and its place has been taken by a circular enamel filling. Eventually, this idea too was dropped, the double trumpets were straightened out, and they begin a new life-style in pairs as represented on a number of objects, mostly cast, amongst which are the cheek-piece (Fig. 15:5) from Birrens, Dumfriesshire;[40] the bridle-bit from Chesterholm, Northumberland;[41] the ovoid mount from Corbridge, Northumberland;[42] the elongated strap mount from Aldborough, Yorkshire;[43] and a dress-fastener found in the Lochspouts crannog, Ayrshire[44] (Fig. 45:12). The distribution suggests that the style was Brigantian, of post-Invasion date. Finally, the triangular mouth gave way to a rounded 'capped' mouth, an idea first mooted by the makers of the Holcombe mirror handle. The new 'cap' is hemispherical, and it covers the mouth. A good example of the new form is to be seen on the belt-plate from Drumashie, Inverness[45] (Fig. 15:6). These embossed twin-capped trumpets, still conjoined at their thin ends, are wrapped around rivet holes, which presumably take the places of bosses.

We are now in an age of 'styles', and capped trumpets are just one of them. Capped trumpets can be seen on another belt-plate, from York,[46] and on a scabbard chape from Hounslow, Westruther, Berwickshire.[47] But, of more lasting interest is the use of the capped trumpet as heads on some dragonesque brooches, like those from Lakenheath[48] (Fig. 48:3), and from Brough-under-Stainmore, Westmorland (Fig. 48:13). Other capped trumpets make up the roundel on the Lakenheath brooch, an arrangement that has more than a passing resemblance to another on the Thames bridle-bit. But the translation from repoussé work to casting tends to lessen sensitivity of treatment.

All the changes and styles noted above occurred in the second half of the first century AD, and in some cases in the early part of the second century. This information is based on some known facts: the Holcombe mirror belongs to the middle of the first century AD; the enamelled bridle-bit from the Thames, with its roundel of trumpets set in the middle of a Greek cross which has been decorated with lozenge patterns, also belongs to the first century, and the lozenges suggest the post-Invasion period; the Lakenheath brooch is considered by some to belong to a federate Icenian context of *c*. AD 40–60; whilst the Birrens cheek-piece came from Level II on Site VIII, and is therefore dated by the pottery to the period AD 122–58.

In summary, it can be said that the continental trumpet motif was first introduced into Galloway at the latter end of the first century BC, and that essentially it became part of a northern style, at the same time fathering some sub-styles. Its adoption by the Brigantes prolonged its history, particularly in view of its appearance on some dragonesque brooches; but, as a head, the motif became debased. This is probably an indication that its purport had been forgotten.

(d) The Cornucopia Motif

The form of the cornucopia today is well known, and in ancient times it was very little different. An example extremely like the modern image of it makes up part of the little silver brooch from Housesteads (Fig. 15:7). The representation here is wholly accurate, even down to the 'pimple' that occurs below the mouth: but, instead of its overflowing with flowers, fruit and corn, two lotus-buds spring from its mouth. For too long the cornucopia has been confused with the trumpet motif, and the cause of the confusion is the bell-shaped mouth. However, an easy method of identification presents itself: a short twist to the thin end is always present in every cornucopia representation.

The Housesteads brooch has been wrongly compared with the Newstead buckle, and then with a piece of open metalwork found at Traprain Law.[49] The error is one of interpretation. It is fortunate that both the cornucopia[50] motif and the lotus-bud motif appear in association on the Housesteads brooch, since this makes for easy recognition. The silversmith did a particularly fine job here, being meticulous in his representation of both motifs, fashioning them into flowing curves. His interest in these motifs was shared by a bronzesmith who made a very striking mount (Fig. 15:8) which was found at Icklingham, Suffolk.[51] This striking and well designed circular mount is made up solely of six cornucopias, occurring here in two sets of three each. The motifs are conjoined at their mouths. The two sets are disposed in triskele fashion, in a bronze openwork design that makes up into a circular object, an arrangement that is very satisfying to the eye, and shows a strong sense of symmetry.

Another openwork mount, which was found in the Roman fort of Arbeia, South Shields,[52] is made up of trumpets and cornucopias. There are twin continental trumpet motifs and three cornucopias. Though a nice piece, it has not the bold simplicity of the Icklingham mount, and therefore its impact is lessened. The Arbeia mount is almost certainly continental and the compilers of the catalogue cited claim northern Gaul or the Rhineland as possible places of manufacture. Lastly, the little brooch (Fig. 15:9) found at Castlelaw Fort, Midlothian,[53] is almost certainly of British manufacture. Three cornucopias and one lotus-bud make up the design.

There is a useful rule of thumb for distinguishing cornucopias from trumpets, with which they are normally confused: when linked together, cornucopias are conjoined at the mouth-ends, whereas trumpets are conjoined at the thin (or mouthpiece) ends.

Otherwise, when occurring singly, as stated above, a twist to the tail will help to distinguish the cornucopia from the trumpet.

(e) The Cable Motif

The cable motif is another which is often mistaken for the trumpet motif; and, in addition, some have referred to it as 'the comma'. As a motif, it is very common in both British and continental art circles, and it has a long history. It is also the most predictable of motifs.

John Boardman[54] gets the credit for distinguishing it in the first place. He maintains that, as a motif, it was introduced to the west as a result of the importation of eastern products in bronze and in ivory. There is a wide range of cable patterns; but a truly European translation of the eastern pattern can be seen in Fig. 16:1, which is to be found included amongst the decoration on a bucket from Moritzing, Tirol.[55] Multiples of the motif occur on this bucket, forming a band round it. But, in the case of the gold open-work-over-bronze-backing found in one of the princely graves at Schwarzenbach,[56] the motif is part of a design of cut-out patterns, composed of slight repoussé tooling and voids, together forming an essential part of the overall design, the whole being a peculiarity of Rhineland artisans. The design is representative of variations played on a number of classical themes, one of the most obvious and recurrent being a pattern composed of two conjoined cable motifs (Fig. 16:2). Including the beaded edges, in all respects the form is very simple, and almost identical with that on the Moritzing cista. Simplicity is the keynote of all representations of this motif, and its predictable form alters hardly at all.

This late-fifth-century-BC goldwork is notable for sophisticated pattern-making. The next example shown (Fig. 16:3) has been taken from a bronze flagon, of early la Tène date, on which stylistic virtuosity plays a major role. This flagon came from a chariot-grave at Durrenberg, Hallein.[57] At the extremity of the handle there is a human head, above which are cable motifs in opposition (Fig. 16:3). This jug is probably early-fourth-century work, and note should be made of the translation of style from slight repoussé to casting. The motif is prominent on the brooch from Ostheim, Germany,[58] a contemporary production, where it flanks both sides of the main curves of the bow, and then appears again on both head and base. Cable motifs are not uncommon on fibulae of the late fifth to early fourth centuries BC. If, as some maintain, the cable motif represents strength, then some literal meaning must be read into its occurrence on these brooches.

Not all continental workmanship was first class. Some, as with the British and the Irish, could be indifferent on occasions, and of such is the badly executed design on a horn-cap (Fig. 16:5) from La Borvandeau, Somme-Tourbe, Marne.[59] This is chased work of the fifth century, and the two cable motifs depicted have tenuous spiral ends. They compare badly with the cast or repoussé specimens already examined. The Marne region is supposed to have been the springboard for much of the art that is found in Britain. Curiously, the cable motif took a circuitous route in reaching Britain, for the next illustration (Fig. 16:6) shows it as it appears on the Turoe stone, in Co. Galway, Ireland.[60] This shaped and rounded stone pillar is a repository of well-known motifs, among which the cable motif figures largely. Another is the lotus-bud motif, and here is an instance in which both appear together, in the same overall decoration. This association will continue, as will be seen later. Although the design here is one suited for carving on to stone, the general appearance of the motif is still very similar to its representations upon metal. Perhaps the stonemason was also a metalworker. On the Torrs pony-cap the motif is extraordinarily similar. The Turoe stone serves to focus attention on what must have been a very common art style in the west of Ireland at this time; yet the remoteness of Co. Galway causes wonder that proficient artists, who were well in touch with current continental styles, should have settled here. It is hard to imagine a la Tène period settlement here, yet it looks possible, in view of the discovery of the Waldalgesheim style gold torc not far away at Clonmacnoise.[61] This torc has buffer-ended terminals, and it is regarded as being of third century BC date. The Turoe stone is not so early. The style of its decoration, incorporating lotus-bud and cable motifs, suggests a date at least two centuries later. The stone-carving technique used is very similar to that employed for the decoration of the carved sandstone pillar from Pfalzfeld, Germany,[62] even though the patterns differ. This and another stone pillar at Irlich, near Neuwied, are domed, and the stonemasons here could have been under the same influences as those who shaped the Turoe stone. The decorative styles on these German pillar stones are not far removed from that current in the late fifth to early fourth centuries, even to that of Waldalgesheim itself. There are no other analogies to go upon, but it is now felt that the Turoe stone must be later than any of these examples, particularly in view of the occurrence round its base of a step-pattern, which has a parallel on the first-century-BC Grimthorpe shield boss.

With alacrity, British artists took to the cable motif. But objects recovered in Britain may not all have been British made, one doubtful specimen being the bronze shield found in the River Witham,[63] a detail from which is given in Fig. 16:8. The decoration is vigorously repoussé, and it has been very well carried out. This detail shows two well executed representations of the cable motif. The style is not late, neither is it very early; but it could antedate the Broighter torc. Strangely, this style of decoration on the terminal discs differs greatly from the style of the decoration on the central boss, which is of an odd conception, though based for the most part on two representations of twin cable motifs conjoined. Here there are the beginnings of a feature which is very characteristic of the Broighter torc, and that is the exaggerated snail-shell-like roundels (boss is hardly the word) that almost obliterate the cable motifs. This coil takes the place of the normal boss, and at Broighter it excites most attention, for it is obtrusive. The decoration on the Witham shield has been described as a rag-bag of motifs.[64] Those represented include the spherical triangle, the cable motif and the palmette, the last named being confined to what has been described as a representation of a bull's head, with staring eyes and with ears flapping. The ears are in the form of split palmettes. Another palmette is situated just above the bridge of the nose. Prolonged contemplation of this overall design will convince of its extreme simplicity, and of its unique nature in Britain; but the basic layout has certain affinities with that on one of the shield bosses from the Thames at Wandsworth.[65] However, the Witham shield boss is earlier, but by how much is not clear. The style is definitely continental, so that the shield may have been made outside Britain. The palmette-shaped ears of the beast, with another palmette below the staring eyes, and the wavy line enclosing the decoration on the roundel, are all continental in character; whilst the hollowed petals of the split palmettes have parallels on the Campanian bucket from Waldalgesheim,[66] and again amongst the silver ornament on a cuirass of the second Kurgan of the Seven Brothers mound, dated between 450 and 440 BC,[67] where the influence is Greek. The wavy line appears on objects like the phalera from Écury-sur-Coole, Marne.[68] This shield is more likely to be of third century BC than of later date.

In its relationship to the present study, of importance is the liberal use that has been made of the cable motif, and of the conservative attitude to it on the Continent; whilst after its adoption in Britain and in Ireland modifications are fairly common, leading to some eccentricities such as the snail-shell-like inturned end but arriving eventually, in the course of another two centuries, at the fully developed boss style.

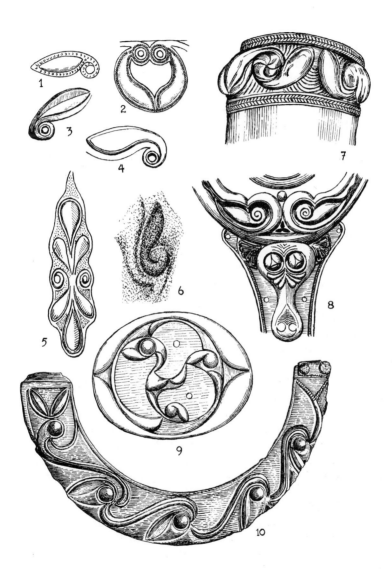

Fig. 16: Cable Motif Decorative Details from: 1. Moritzing, Tirol. 2. Schwarzenbach, Germany (5th Century). 3. Durrenberg, near Hallein. 4. Ostheim, Germany. 5. La Borvandeau, Somme-Tourbe, Marne. 6. Turoe Stone, Co. Galway. 7. Gold Torc, Snettisham, Norfolk. 8. Bronze Shield, River Witham. 9. Bronze Plaque, Llyn Cerrig Bach, Anglesey. 10. Bronze Collar, Llandyssul, Cardigan. 9 and 10 (1/2) remainder not to scale

There was thus no long jump forward to a boss style in Britain; it was just the result of slow evolution. Leeds looked upon the boss style as being regional to the north. There are some close examples from Wales, but with local differences. The bronze plaque from Llyn Cerrig Bach[69] (Fig. 16:9) has decoration which is a good example of the Welsh style, here in repoussé. The basis of the design is the cable motif, in which the inturned ends, snail-shell-like in Ireland, have here been developed into bosses. Elongated forms of the motif, conjoined, enclose the whole design. The general effect is one of an embryo boss-and-petal pattern, and one feels that the real boss style is not far away. This Llyn Cerrig plaque is probably of first century BC date, and the decorative style here is carried a step forward, and with more assurance, to a repetitive running design on the half-collar from Llandyssul, Cardiganshire.[70] The decoration here (Fig. 16:10) shows running motifs linked by S-scrolls. The bosses have been applied separately: they consist of studs with hemispherical heads, and they have been driven into the bronze plate and then riveted over at the back. It was a time-saving idea, and one that was applied to the Lochar Moss beaded collar, on which the broken back scrolls were riveted on in the same manner. There is a strong possibility of technical exchanges between the two areas having taken place, leading to an interchange of ideas, relative to decoration and so on.

(f) The Spherical Triangle

This motif can be described as a triangle with concave sides; or it can be compared with the void that is left at the centre by three touching spheres – hence its name. The motif is of continental origin, and it was first observed on the bronze mount from Commachio (Fig. 12:1a). It was not of common occurrence, like the last motif, and it appears to have been somewhat aloof from the decoration amongst which it appears. The spherical triangle occurs on the bronze bracelet from Tarn,[71] on which it is enclosed within a domed roundel against a roughened background (Fig. 17:1). The umbonate nature of the roundel and its separation emphasise this isolation of the motif. In this particular form the motif was translated into British art.

The spherical triangle appears to have been of wider occurrence in Britain than it was on the Continent. On the Tarn bracelet it is deeply channelled into the metal, and so it is in its occurrence on the bridle-bit from the Thames at London (Fig. 15:4), and again on snaffle-bits from Ringstead, Norfolk.[72] In all cases the motif is enclosed within a roundel of umbonate form, and generally large at that. But there was a change to the repoussé style, examples of which can be found on the mirror mount (Fig. 5:3) from Balmaclellan, Kirkcudbright.[73] The method employed for its representation on the central boss of the Witham shield is rather curious: for here the motif is composed of three seed-shaped pieces of a material that looks like coral, which are kept in place with pins. There is a resulting central void, which is of the spherical triangle shape. If one looks at the Tarn spherical triangle it will be noted that these seed-shaped parts are there, with marks made by a hollow punch at their middles; and this representation may have suggested the shorter cut of pinning seed-shaped pieces of metal to a flat surface, with virtually the same effect.

Linear representations of the spherical triangle are common, and generally they are much later. Chasing was probably the technique used, as it was with the mirror decoration. On the reverse sides of some mirrors there may be a spherical triangle, or maybe two, hidden away somewhere amongst the decoration. But it will be isolated from the

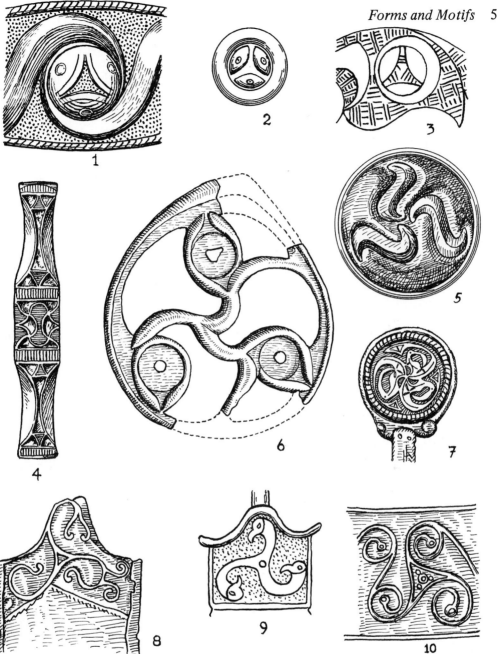

Fig. 17: Spherical Triangle and Triskele Motifs 1. Decorative Detail on Armlet, Tarn, France. 2. Detail from Shield Boss, River Witham. 3. Decorative Detail on Bronze Mirror, Trelan Bahow, St Keverne, Cornwall. 4. Enamelled Cheek-Piece, Polden Hill, Somerset. 5. Decorative Detail on Silver Disc, Manerbio sul Mella, Brescia, Italy. 6. Openwork Bronze Disc, no Locality. 7. Pinhead, Tully Townland, Co. Antrim. 8. Decoration on Sword Scabbard, La Tène. 9. Decoration on Sword Scabbard, Obermenzingen, S. Germany. 10. Decorative Detail, from Gilt-Bronze Helmet, Ampreville, Eure, France. (Not to scale, except 6 (3/4))

rest of the pattern, and this is achieved by enclosing it within a circle. The example shown in Fig. 17:3 has been taken from the Trelan Bahow mirror, St Keverne, Cornwall.[74] The motif occurs twice on the Birdlip mirror (Fig. 7:1), and on the Colchester mirror it occupies a central position. There are two occurrences on the Desborough mirror. It will be seen that the tendency is to hide it away; whereas on continental objects it occupies pride of place. It could have been a talisman. This is suggested by its occurrence on fighting equipment, such as the Hunsbury and Meare scabbards (Fig. 32:1 and 2). The motif was brought into the home on pottery, such as the Iron Age ware found at Meare, Somerset.[75] Horsemen had it incorporated into the decoration on cheek-pieces.[76]

(g) The Triskele

As a motif, some might regard the triskele as a doubtful starter; but as a constituent feature of many a composition, it is perhaps more dominant than some of the other motifs considered above.

Triskeles have three arms springing from a common centre, which is either triangular or it is marked by a small circle or boss. Both of these forms are of continental origin. Triskeles of the first type form the basis of the design on the thin foil-covered disc from Schwarzenbach.[77] They are prominent on the Tarn bracelets, on which they occur in plastic form,[78] and here they are of third-century date. A Manx-like triskele is the sole decoration on the central hollow boss on a silver disc, of the same period, which comes from Manerbio sul Mella, Brescia, in north Italy[79] (Fig. 17:5). This third-century repoussé style, with its up-ended arms, could have been amongst the ancestral types of the unlocalised open-work bronze disc in the Ashmolean Museum[80] (Fig. 17:6); though here there is an emergent boss-and-petal style superimposed upon it. With its peculiar broken-back curves, which Fox thought of as imitative of the branches of a tree trunk, this piece could conceivably have set the pattern for other works of Celtic art, notably the Icklingham mount. The arrangement never ceased to occupy the minds of Celtic craftsmen, and there was a persistence of it in Ireland, until late into the heroic age, as evidence of which there is the pin found associated with 'souterrain ware' in a ring-fort in Tully Townland, Co. Antrim[81] (Fig. 17:7).

Whereas triskeles of the first type invariably have arms that curve to the left, the second type has arms that normally curve to the right. The rule does not always hold good, since occasionally there is some uncertainty or confusion. The second type is more common on the Continent: typical occurrences are on a la Tène sword scabbard (Fig. 17:8) and on a fourth-century-BC gilt bronze helmet (Fig. 17:10). The sword scabbard comes from la Tène, whilst the helmet comes from Ampreville, Eure, France.[82] The triskele on the second-century-BC scabbard (Fig. 17:9) from Obermenzingan,[83] could be of either type: its arms curve to the left. The ornithomorphic finials are interesting in another light, for they can be compared with others on the Wandsworth boss.[84] In place of the central triangle there is a circle, also to be seen on one of the phalerae from Écury-sur-Coole, Marne.[85]

Two mounts from Tal-y-llyn, Merioneth,[86] each have a triskele on which a small knot or rivet occupies the central position. One triskele is defined only by a well-executed basketry background: the other triskele is in the repoussé style. But both are identical, with comma tails to the arms. Perhaps their form foreshadows the coming of the more complex triskele on the Moel Hiraddug bronze plaque,[87] a well-developed example of local craftsmanship. The elongated tails to the arms have lotus-bud motif overtones. The Tal-y-llyn mounts probably belong to the late first century BC, but the Moel Hiraddug plaque is definitely of first-century-AD manufacture.

This Welsh style differs greatly from that of north Britain. There is a boldness of approach to works of this sort, with which Brigantian productions do not compare very well. Up till now, triskeles generally have been plain; but the Brigantes thought fit to add to the basic form, so the tail ends become spiralled. This idea caught on, and broken spirals based on this idea can be seen on the little disc-brooch from Brough, Cumbria[88] (Fig. 4:4). Actually, the spiral is made up of a petal and a tendril, the tendril forming the spiral. In this there are lotus-bud overtones, and confirmation of this belief can be seen on the Silchester disc-brooch (Fig. 4:2), where the finials are of the same form as are the lotus-buds on the Aesica brooch. These disc-brooch patterns must therefore be related to the Aesica style, in which lotus buds and continental trumpets are included as part of that style.

Triskeles are the foundation, or the basis for further decoration: but they can be represented as they are without further additions. The triskeles on these small disc-brooches do not conform either to one type or to the other. This lack of differentiation is noted only in Britain. As a motif, it survived the Occupation, chiefly in the fringe areas and in Ireland. Many an occurrence is in openwork, a striking example of which is the representation on the late-first-century tankard handle from Trawsfynydd,[89] and somewhat later in multiple openwork form on the scabbard mount, associated with second-century dolphin brooches, from Lambay Island.[90] In a specialist cast style the triskele forms the basis of the design on the Longban Island, River Bann disc[91] (Fig. 35:2). The Lambay Island style was later copied, but this time against an enamel background, as it appears on the escutcheon of the Kingston hanging-bowl (Fig. 81:1). The layouts of the spiral patterns on the hanging-bowl escutcheons and prints have a basic triskele form. In Ireland, the triskele took on an entirely new personality when elongated trumpets were used in its representation, instead of plain lines. The Bann disc illustrates this point very well.

NOTES

1. Paul Jacobsthal, *Early Celtic Art* (Oxford University Press, 1969), p. 60.

2. *BM Guide to Early Iron Age Antiquities*, p. 19, Fig. 9.

3. E.M. Jope, 'The Waldalgesheim Master' in *The European Community in Later Prehistory* (Routledge, 1971), pl. 22.

4. As on the crescent-shaped mount from Brunn-am-Steinfeld, Austria.

5. Cyril Fox's words, in the course of his comment on the design on the bronze shield mount from St Mawgan-in-Pyder, in *Arch. Journ.*, CXIII (1956), pp. 33ff.

6. From Commachio, Italy, and now in the Museum fur Vor-und-Fruhgeshichte, Berlin. *Praehistoriche Zeitschrift*, 25 (1934), pp. 62ff.

7. Jacobsthal[1], pl. 247a.

8. Ibid., p. 94.

9. L. Murray Threipland, 'The Excavation at St

Mawgan-in-Pyder, north Cornwall', *Arch. Journ.*, CXIII (1956), p. 80, pl. XI, and Fig. 40.

10. Fox[5], p. 81.

11. J.W. Brailsford, 'The Polden Hill Hoard, Somerset', *Proc. Preh. Soc.*, 41 (1975), p. 222.

12. This type comes into being about the middle of the first century, and it continues until its end.

13. Brailsford[11], pl. XXII.

14. Perhaps the style was introduced by a visiting Greek. A Greek schoolmaster from Tarsus was able to undertake a sight-seeing trip to the Western Isles of Scotland in AD 82.

15. M. Duignan, 'The Turoe Stone: Its Place in Insular la Tène Art' in P.-M. Duval and C. Hawkes (eds.), *L'Art Celtique en Europe protohistorique . . .* (1976), p. 201.

16. R.J.C. Atkinson and S. Piggott, 'The Torrs Chamfrein', *Arch.*, XCVI (1955), pp. 197ff.

17. E.C.R. Armstrong, 'The la Tène Period in Ireland', *Journ. Roy. Soc. Antiq. Ireland*, LIII (1923), p. 14, Fig. 9.

18. F. Henry, *Irish Art* (Methuen, 1947), pl. 7a.

19. C. Fox, *Archaeology of the Cambridge Region* (1923), p. 81, pls. XV:5; XVIII:2x.

29. Atkinson and Piggott[16], p. 228.

21. E.M. Jope, 'An Iron Age Decorated Sword Scabbard from the River Bann at Toome', *Ulster Journ. Arch.*, XVII (1954), p. 81.

22. Curle, *A Roman Frontier Post and its People*, pl. LXXVI:2.

23. E. Birley and J. Charlton, 'Third Report on Excavations at Housesteads', *Arch Ael.*[4], XI (1934), p. 197, pl. XXIXe:81. See also Ibid., XXXIX (1961), p. 36, pl. VII:5.

24. V. Gordon Childe, 'Excavations at Castlelaw Fort, Midlothian', *Proc. Soc. Antiq. Scot.*, LXVII (1932–3), p. 385, Fig. 13:2.

25. J. Collingwood Bruce, *Catalogue of Antiquities at Lanwick Castle*, p. 146, no. 791.

26. Jacobsthal[1], p. 19.

27. S. Piggott, *Ancient Europe* (Edinburgh University Press, 1973), p. 243, Fig. 135.

28. R.A. Smith, 'Two early British Bronze Bowls', *Antiq. Journ.* VI (1926), pp. 276-83.

29. Jacobsthal[1].

30. T.G.E. Powell, *Prehistoric Art* (Thames and Hudson, 1966), p. 204.

31. R.A. Smith, 'Specimens from the Layton Collection in Brentford Public Libraries', *Arch.*, LXIX (1920), p. 22, Fig. 22.

32. A.J. Evans, 'On a late Celtic Urnfield', *Arch.*, LII (1890), p. 360, Fig. 11.

33. E.T. Leeds, 'An Enamelled Bowl from Baginton, Warwickshire', *Antiq. Journ.*, XV (1935), p. 109.

34. H.E. Kilbride-Jones, 'A Bronze Bowl from Castle Tioram, Moidart; and a suggested absolute Chronology for British Hanging-Bowls', *Proc. Soc. Antiq. Scot.*, LXXI (1936–7), p. 207, Fig. 2:2.

35. Ibid., Fig. 1, and Fig. 2:1.

36. J.P. Bushe-Fox, 'Second Report of the Excavation of the Roman Fort at Richborough, Kent', *Soc. Ant. Research Report*, no. VII, p. 46, pl. XIX:33.

37. D. Wilson, *Prehistoric Annals of Scotland* (Sutherland & Knox, 1851), II, p. 146, Fig. 131. See also *Proc. Soc. Antiq. Scot.*, XV (1881), p. 351, Fig. 28.

38. A. Fox and S. Pollard, 'A decorated Bronze Mirror from an Iron Age Settlement at Holcombe, near Uplyme, Devon', *Antiq. Journ.*, LIII (1973), p. 26, Fig. 5.

39. C.H. Read, in *Proc. Soc. Antiq.*[2], XXIII (1909–10), p. 159.

40. Eric Birley *et al.*, 'Excavations at Birrens, 1936–37', *Proc. Soc. Antiq. Scot.*, LXXII (1937–8), p. 337, Fig. 38.

41. J.D. Cowen, 'A Cheek Ring of a Celtic Bit', *Proc. Soc. Antiq. Newcastle.*[4], VI (1933–4), p. 223.

42. M. Macgregor, *Early Celtic Art in North Britain* (Leicester University Press, 1976), no. 15.

43. Ibid., no. 30.

44. R. Munro, *Lake Dwellings of Europe* (Cassell & Co., 1890), p. 423, Fig. 153.

45. This belt-plate was found in association with the dress-fastener shown in Fig. 45:9. See *Proc. Soc. Antiq. Scot.*, LVIII (1923–4), p. 11. Fig. 1.

46. C. Fox, *Pattern and Purpose* (National Museum of Wales, 1958), p. 119, pl. 52:C.

47. J.N. Graham Ritchie and A. Ritchie, *Edinburgh and S.E. Scotland: Regional Archaeologies* (Heinemann Educational, 1972), p. 49, Fig. 33a.

48. R.R. Clarke, 'The Iron Age in Norfolk and Suffolk', *Arch. Journ.*, XCVI (1939), p. 55, Fig. 10:1. Fox[46], pl. 41:b.

49. D. Charlesworth, 'Roman Jewellery found in Northumberland and Durham', *Arch. Ael.*[4], XXXIX (1961), p. 36, pl. VIII:5.

50. Cornu-copiae, late Latin for the horn of plenty, which sometimes figures in Roman art. Originally, it was the horn of the goat Amalthea, by which Zeus was suckled.

51. *BM Guide to Antiquities of Roman Britain*, p. 28, Fig. 14:5.

52. Piggott, *Early Celtic Art*: a catalogue of an Exhibition sponsored by the Edinburgh Festival Society (1970), pp. 32, 166.

53. Gordon Childe[24], p. 385, Fig. 13:2.

54. Boardman, 'A Southern View of Situla Art' in Boardman (ed.), *The European Community in later Prehistory*, p. 130.

55. Jacobsthal[1], p. 204, no. 399.

56. Ibid., p. 167, no. 18.

57. Ibid., p. 201, no. 382.

58. Ibid., p. 194, no. 315.

59. Ibid., p. 185, no. 168.

60. Duignan[15], Figs. 4 and 5.

61. Armstrong[17], Fig. 9.

62. Jacobsthal[1], p. 165, no. 11.

63. The best illustration is in J.V.S. Megaw, *Art of the European Iron Age* (Adams & Dart, 1970), p. 149, no. 252.

64. Ibid., p. 150.

65. *BM Guide to the later Prehistoric Antiquities of the British Isles*, p. 70, pl. XX:2.

66. Jacobsthal[1].

67. Ibid., p. 30, pl. 221:f.

68. Ibid., p. 187, no. 189.

69. C. Fox, *A Find of the Early Iron Age from Llyn Cerrig Bach, Anglesey* (National Museum of Wales, 1946), pl. XVIII.

70. *National Museum of Wales Guide to the Collection Illustrating the Prehistory of Wales*, pl. VIII:2, no. 464.

71. Jacobsthal[1], p. 275, pl. 150.

72. R.R. Clarke, 'A Hoard of Metalwork of the early Iron Age from Ringstead, Norfolk', *Proc. Preh. Soc.*, XVII, pp. 214ff.

73. J. Anderson, *Scotland in Pagan Times; the Iron Age* (David Douglas, 1883), p. 126, Fig. 104.

74. *BM Guide*[2], p. 121, Fig. 132.

75. A. Bulleid and H. St George Gray, *The Glastonbury Lake Village*, II, pp. 209, 212.

76. *BM Guide*[2], p. 144, Fig. 163.

77. Jacobsthal[1], p. 169, no. 34.

78. Ibid., p. 192, no. 275.

79. Ibid.

80. C. Fox, 'An Openwork Bronze Disc in the Ashmolean Museum', *Antiq. Journ.*, XXVII (1947), p. 1, pl. I.

81. Unpublished. Illustrated here by courtesy of Alan Harper.

82. Jacobsthal[1], p. 140, pl. 79.

83. J.M. de Navarro, 'A Doctor's Grave in the middle la Tène Period, from Bavaria', *Proc. Preh. Soc.*, XXI (1955), p. 231.

84. *BM Guide*[65], pl. XX:1.

85. Jacobsthal[1], p. 189, pl. 117.

86. Megaw[63], p. 156, no. 262; p. 158, no. 267.

87. *National Museum of Wales Guide*[70], pl. VII.

88. *Proc. Soc. Antiq.*[2], XIX (1901–3), p. 130.

89. J.X.W.P. Corcoran, 'Tankards and Tankard Handles of the British Early Iron Age', *Proc. Preh. Soc.*, XVIII (1952), p. 97.

90. R.A.S. Macalister, 'On some Antiquities found on Lambay', *Proc. Soc. Irish Acad.*, XXXVIII C, p. 243, pl. XXIV; E. Rynne, 'The la Tène and Roman Finds from Lambay, Co. Dublin: a re-assessment', *Proc. Roy. Irish Acad.*, 76 C (1976), pp. 231ff.

91. E.M. Jope, 'The decorative Cast Bronze Disc from the River Bann, near Coleraine', *Ulster Journ Arch.*, XX (1957), p. 95.

4. Patterns

Whereas motifs are specifically simple, and are the constituent features or the dominant ideas in artistic compositions, patterns are decorative designs normally composed of more than a single idea. Alternatively, patterns can be composed of a number of linked motifs. Motifs are widespread, whereas patterns may be localised. Of course, patterns can be simple too; or they can be very complex. Examples of a simple pattern are the Sunburst or the Boss and Petal. In the former there are two elements, the sun and the sun's rays; in the latter there are the boss and the petal. Whereas motifs are entirely self-contained, patterns, such as the chevron pattern, can go on for ever. These are some of the essential points by which motifs can be distinguished from patterns, and they are set down here to explain the division between one and the other.

Several new patterns were reaching Britain in the first century AD, and the majority were alien to the British mind. But the Celt, not being an originator himself, nevertheless had an appetite for new forms, so that many of these new patterns were introduced into the art during the early Romano-British period. Mostly, they were provincial Roman patterns, and good examples are to be seen on the bronze bowl from Snailwell, Cambridgeshire.[1] This bowl was imported from Gaul: with it were found representative Gaulish pottery forms, very similar to pre-Conquest pottery at Colchester; and the whole lot was buried with the remains of a British warrior, who had died a year or two after the Claudian Invasion of AD 43. Inside this bowl there are two patterns – the sunburst and the chevron – both of which were soon assimilated into British art, even to the exclusion of more ancient, and therefore Celtic, patterns or motifs. This was happening everywhere during the early years of the Occupation.

(a) The Sunburst Pattern

This pattern was introduced into Britain on the inner base of the Snailwell bowl. The pattern (Fig. 18:5) consists of a number of loops springing at right angles from a central roundel or orb, at the centre of which there is a pattern which can be variously interpreted, but which has some resemblance to a swastika. The whole is enclosed within a border of chevron pattern. Clearly, the composition is meant to be representative of the sun and its rays; and for this reason it has been named 'the sunburst pattern'.

The early adoption by the British metalworkers of this alien sunburst pattern, and its many appearances on dress-fasteners and disc-brooches, may have been for the purpose of pleasing dedicated sun-worshippers (in the religious, not the modern sense) but also with an eye to business. There is a very good, almost modern representation of the sunburst pattern on a dress-fastener (Fig. 45:6) which was found in the Roman fort at

Mumrills.[2] This dress-fastener must have been lost some time during the brief occupation of the fort during the Agricolan campaign in AD 80 or 81. It is a circular or disc-fastener, and the disc has a milled edge: the pattern on the disc consists of a fiery orb, from which long and carefully graduated flames shoot out at right angles. The centre of the orb is occupied by a quatrefoil pattern,[3] giving to this piece the stamp of British authenticity. The pattern is drawn against a background of enamel, which is alternately red and yellow, whilst blue forms the background to the quatrefoil pattern. Altogether, the workmanship is first class.

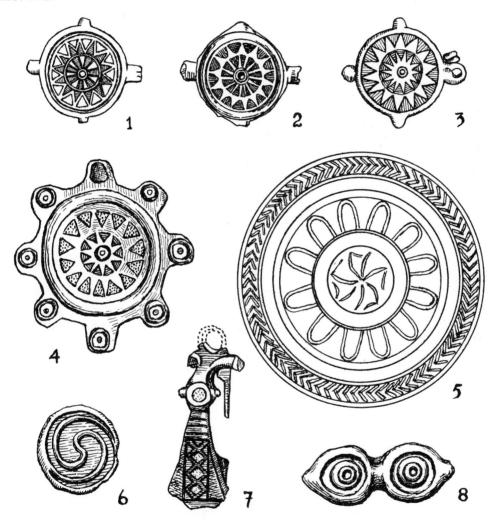

Fig. 18: Umbonate enamelled Disc-Brooches 1. Traprain Law (Red, White and Blue). 2. Nor'nour, Isles of Scilly (Dark Blue, Light Blue). 3. Alchester (Red, Green and Blue). 4. Disc-Brooch, enamelled, Hilderley Wood, Malton, N.R. Yorkshire (Red, Blue). 5. Decorative Detail from Bowl, Snailwell, Cambs. 6. Button, Corstopitum, Northumberland. 7. Fan-Tail Brooch, Traprain Law. 8. Double Boss-and-Petal Design Mount, Corbridge. (1/1) except 4 (3/2)

But, because the sunburst was an alien pattern, perforce it had to be modified. As a result of those modifications, there came into being a thoroughly geometric design. The flames became like stretched out lozenges, and even the quatrefoil was treated in a like manner. This modified style appears on a dress-fastener from Slack, Yorkshire (Fig. 45:7). Roughly at the same time, the new modified version of the sunburst pattern was applied to umbonate disc-brooches, but a final modification resulted in the tongues of flame being represented by triangles, whilst the quatrefoil pattern is replaced by a star. The design is now thoroughly geometric, and henceforth it alters hardly at all, being repeated again and again from the Scillies[4] to Traprain Law.[5] The design is enamel-filled, usually with alternate colours. Variations were more likely to apply to the metalwork, as in the case of the Hilderley Wood, north Riding of Yorkshire[6] brooch, patterns generally remaining the same.

Fabrication occurred mostly in the second century, and the closest dated specimen, Fig. 18:3, comes from Alchester.[7] This brooch was found in metalling between sites A and B in Period B, during which the occupation began in the reign of Domitian, and ended *c*. AD 130. This date accords well with the Traprain Law evidence, for Fig. 18:1 came from the lowest level, implying a late-first-century, or a very-early-second-century date for this item. The Nor'nour brooch shown (Fig. 18:2) belongs to the second century. Confirmation of these dates is provided by brooches from Camelon[8] (AD 80–180) and from Newstead,[9] at which fort a specimen was found in the baths, in a context dated to the early/mid second century. The Wroxeter[10] evidence gives an earlier date, for here a brooch was found in association with coins of Vespasian. Other brooches were found at Corbridge, Caistor by Norwich, Colchester, Carlisle, Harlow and Richborough.[11] The high incidence of discovery on Roman sites is a measure of the popularity of the brooches with the Roman soldiery, and this fact might explain the adoption by the metalworkers of an alien pattern like the sunburst.

The daisy-like pattern on the dress-fastener (Fig. 45:8) from Caerleon,[12] Monmouth, could have come into being only as a result of a further simplification of the sunburst pattern. Although this specimen was found in south Wales, the form is more likely to have been northern, perhaps Brigantian in origin, one exactly similar having come from Catterick, with red rays on a yellow background.[13]

The pattern appears nowhere but on personal items like dress-fasteners and umbonate disc-brooches, in the series of bronze objects made in Britain during the first and second centuries.

(b) The Lozenge Pattern

Lozenges appear in multiples. British metalworkers may first have learned about them from their appearances on small objects like fan-tail brooches. One of these managed to get to Traprain Law,[14] where it was found in the lowest level. It is a Claudian brooch (Fig. 18:7) and it is a much travelled specimen. There is another exactly like it from Colchester,[15] where it was found in a first-century context.

An instance of early British usage of this pattern, in the first century, is to be found on the central design on the bridle-bit from the Thames at London (Fig. 15:4), on which lozenges appear on the Greek cross-shaped central plate. The Brigantes also found lozenges to be attractive. They made extensive use of them as decoration for the central panels of their dragonesque brooches. The popularity of the pattern amongst the British was such that it was taken to Ireland, where it was included in the repertory of designs kept by a metalworker at Dooey, Co. Donegal[16] (Fig. 67:1a). The pattern also occurs on Irish zoomorphic brooches (Fig. 71:2). The Irish tended to reduce the size of the lozenges, though two sizes, one of normal dimensions, the other smaller, are represented on the latchet from Castle Island, Dowris, Birr, Co. Offaly[17] (Fig. 67:5).

(c) Spirals

Spirals are of extreme antiquity. Linked, they are found on Danubian pottery; on ware from Dikilitash; at Drachmani; at Naxos; and on a stone stele belonging to the sixteenth-century-BC Mycenean culture.[18] Spirals decorate the walls of some of the Maltese temples. The Celts were perhaps a little slow in adopting spirals, for they do not figure largely in continental la Tène art, nor were they common in pre-Conquest Britain. Some of the continental examples can be quoted here: spirals appear on the gold-covered disc from Auvers-sur-Oise, France,[19] which is of fourth-century-BC date; and on bronze torcs from Prosuis, Marne.[20] But, because spirals are essentially curvilinear, they have always been looked upon as mainly Celtic.

Spirals come in three forms: single, double and triple. Single spirals are easy to fashion, since they are nothing more than a diminishing circle. But there is a single spiral which distinguishes itself from the rest: this is the one with swollen finials. This is essentially an Irish variation, which became common in later years in Ireland; yet, though peculiarly Irish, this type of spiral was first noted at Prunay, Cher,[21] making its initial appearance in Ireland on the Turoe stone.[22] Sometime during the fifth century AD the finials became hollowed; and when the central finial is so treated, it tends to create a false impression, for it can be mistaken for a genuine double spiral. Some of these supposed double spirals appear on zoomorphic penannular brooches, and good examples can be seen on the Dowris latchet (Fig. 67:5). On the disc quoted it will be noted that two of these hollowed finials are entwined, whilst to the right and left others can be seen. Genuine double spirals are not as common as singles or triples; but there is a double spiral on a button-like object from Corstopitum[23] (Fig. 18:6). In Ireland, good examples appear on the Group B zoomorphic penannular brooches from Athlone (Fig. 70:3).

Triple spirals are more common in Britain than they are in Ireland, though this assessment may not be correct, in view of the large amount of Irish material that has been destroyed, or lost. However, good examples can be seen in Fig. 76:3 and 5. The quality of the workmanship is such that there must have been many more of these about at one time. The best triple spiral examples are those on the hanging-bowl escutcheons and prints, and found in Britain mostly to the east of the Fosseway. But triple spirals were also carved on to wood; one example, carved on an ash panel, was found along with head-stud and trumpet brooches of the early second century AD at the Lochlee crannog, Tarbolton, Ayrshire.[24]

(d) Linked Spherical Triangles

Normally, spherical triangles are represented singly. Linked, they are of rare occurrence. However, there are some fragments of tubes split lengthways which had been part of a chariot burial at Somme-Bionne;[25] and here the entire openwork decoration is in the form of spherical triangles, conjoined tip to tip. The arrangement is cleverly carried out.

There is no evidence of the occurrence of multiple spherical triangles in Britain,[26] but in Ireland it was the custom to pair them, though, of course, singles occur as well. The pairing was due to a misunderstanding, for the Irish confused the design on the Knowth brooch (Fig. 66:2), itself obtained from the pattern on the handle of the Nijmegen mirror (Fig. 21:2), with spherical triangles in opposing positions, and this brought about the pairing. For the same reasons, the Irish spherical triangles have dots at their centres.

Paired spherical triangles are common on the Newry latchets (Fig. 66:5), and on zoomorphic penannular brooches of the B Group (Fig. 66:3 and 4). But the fashion had a short history, being mostly of third century date in Ireland: certainly it did not outlast the fourth century.

(e) Boss-and-Petal

This is essentially a British pattern. The pattern is made up of a hemispherical boss set in a hollow, just off-centre on a petal-shaped casting, which itself would have been circular, but for its being pointed on one side. The form is best illustrated by the well-patterned dress-fastener from Colchester, Essex[27] (Fig. 26:3).

There are at least three ways in which the boss-and-petal pattern could have originated. But first, particular note must be taken of the fact that on the point of the petal of the Colchester specimen there is a vestigial line, which starts at the tip and then proceeds longitudinally as far as the hollow. This line is a clue to the origin of the form.

Of the three possible origins, the first is Welsh. On the Llyn Cerrig Bach plaque, and again on the half-collar from Llandyssul (Fig. 16:9 and 10) there is a pattern that can be claimed to be very similar to the boss-and-petal, the basis of which is the cable motif. On these Welsh pieces, the inturned ends of the cable motifs have been modified into a boss form, and these bosses, particularly in the case of those on the Llandyssul collar, are actually hemispherical headed rivets. To eliminate the one-sided effect produced by this modification, a moulding in low relief was added in an effort to balance the design, and this successfully created an illusion. The resulting design looks very like a true boss-and-petal.

Secondly, there is a lesser possibility that the boss-and-petal pattern might have been derived from a pre-Conquest curvilinear design like that (Fig. 26:4) on a heavy bronze plaque with champlevé enamel decoration found at Woodlands, Dorset,[28] which appears to be of early-first-century-AD date. This kind of pattern is not uncommon in the mirror style of decoration: for instance, it occurs on the Desborough mirror, and again on a sword scabbard from Meare, Somerset.[29] And in openwork form, the pattern can be seen on the bronze scabbard mount (Fig. 26:6) from Brough, Cumbria.[30]

The third and most likely origin is at first glance less obvious. Amongst the decorative elements making up the design on the bronze bridle-bit (Fig. 26:1) from Holderness, East Yorkshire,[31] are two opposed roundels which are suspended from the link, both being enamelled in blue with quatrefoil patterns at their centres. Each is contained within a frame composed of overlapping double continental trumpet motifs. These are literally wrapped around the roundels. On the outer sides, where they abut on to the main ring, the pairs meet mouth to mouth; but, on the inner sides, one passes beneath the other. Already, there is a strong suggestion here of the boss-and-petal pattern. Some early dress-fasteners from Traprain Law (Fig. 46:12) are clearly reminiscent of this layout, even to the extent of having a flat-topped 'boss' (here also decorated), which is more like a non-boss. The flattened top is for the purpose of facilitating decoration. Note the trumpet-shaped surround on the Traprain Law dress-fastener. Later, the sharp catching edge of the trumpet mouth was smoothed off by bringing up the thin end of the trumpet in a sort of tail-swallowing act. The smoothing out of this feature results in the formation of a petal-shaped form, the slightly concave line on the Colchester dress-fastener being there as a reminder of its trumpet beginnings.

This simplification of the story of how a trumpet became a petal might be thought fanciful. But it must be remembered that the metalworkers were devising a form that would be suitable for repetitive work in the usual manner of casting from prepared moulds. The new form may have been devised at Traprain Law, a factory centre that turned out hundreds, if not thousands of boss-and-petal dress-fasteners, many for export to other areas. Leeds' claim that the style is northern is sustained. In the near mechanical productivity period of the second century, the boss-and-petal pattern was flogged; but its sudden demise was due to the destruction of the workshops in *c*. AD 196.

(f) The Quatrefoil Pattern

This is a simple pattern made up of four petals set in the form of an equal armed cross. It is included here because it is not an established motif, having no ancestry, and it had no future, because it was quickly forgotten. The pattern seems to have been of Brigantian origin, and it appears quite suddenly in the first century, particularly on the Holderness bridle-bit (Fig. 26:1), on which it forms the centre of the enamel decoration. It is also the main decoration on the terret from Seven Sisters, upper Dulais, Glamorgan (Fig. 25:3). Somewhat later, it appears on two dragonesque brooches (Fig. 51:6 and 7); then it seems to have been phased out.

The pattern travelled to Ireland: it makes a single appearance on one of the zoomorphic penannular brooches, and the metalworker of Loughcrew included it amongst his patterns, preserved for us on the bone trial pieces (Fig. 34:12).

(g) The Chevron Pattern

This pattern could be of provincial Roman origin. Its early appearance on the Snailwell bowl has already been noted. It is of common occurrence on decorated Samian ware of a form that first appeared *c*. AD 70, and its popularity lasted until the third century.[32] In Britain its history would have been straightforward enough but for the chevrons on the Snettisham torcs (Fig. 16:7). The specimen shown here is not often illustrated. On the terminal there is an association of chevrons with double cable motifs, of which two spring from a common boss. This boss is rosette-shaped: there is a useful parallel between this and the rosettes on the Aylesford bucket. This particular gold torc from Snettisham was found with another, which, so far as its decorative style is concerned, can be put along-side the Aesica brooch. Valuable objects such as these invariably enjoyed long lives; so that there is no reason why this torc with the chevron patterns should not have been contemporary with the Aylesford bucket. The Aylesford bucket was associated with a final la Tène brooch, and it was contemporary with the early Augustan period,[33] giving a possible late-first-century-BC date for the bucket. The Belgic Aylesford cista and the Snailwell bowl could both have come from an area within a few miles of each other, the bowl being old when it was buried with the Iron Age warrior in the middle of the first century AD. But, whereas both the Aylesford cista and the Snailwell bowl were imports, the Snettisham gold torcs are thought to have been Irish made.[34] But amongst these, that illustrated in Fig. 16:7 is the odd man out, with its rosettes and chevron patterns, which encircle the terminal in two places. There are deep implications here, which need further examination.

Later, in this book, we shall study the implications of the Elmswell-Broighter style, which influenced decoration from Yorkshire to Derry: then it will be seen that the Snettisham torc belongs also to this style, suggesting that it is out of place in East Anglia. Perhaps it might seem more realistic to regard this torc, with its chevrons and its rosettes, as the work of continental metalworkers. It is a possibility that can neither be proved nor disproved.

Chevron patterns are common in Scotland, where they are to be seen amongst the decoration on the massive armlets (Fig. 9:2 and 3). South of the Wall, chevron patterns appear on a fragmentary scabbard belonging to a sword with a curiously shaped cast bronze hilt and pear-shaped pommel, which comes from Worton, Lancashire.[35] Piggott[36] includes this sword and scabbard within his Group IV, which is Brigantian. But this one differs completely from the other Group IV swords and scabbards; but there is another pommel exactly like this one from Brough, Cumbria,[37] and the Brough specimen also has the chevron pattern down its sides (Fig. 32:6). The fluted caps to the pommels would seem to indicate Roman influence. Curiously, the raised lobes on the pommels are vaguely reminiscent of those placed above the Pfalzfeld heads.[38]

(h) Hexafoil Patterns

Multiples of this motif are not common. They can be seen on a bronze spout flagon from Eygenbilsen,[39] and also on a bronze plaque from Castelletto sopra Ticino.[40] Since hexafoils can be backdated to at least 400 years BC, such carefully executed and well thought-out patterns indicate early use of the compass: for, as Frey has pointed out, all this is compass work, and without the use of a compass, accuracy would be impossible.

Such patterns must have travelled slowly, since they were late in reaching Britain. The Romans had an emasculated version, which is of common appearance on altars, tombstones and antefixes.[41] It was this emasculated version which must have reached Ireland first: it was amongst the patterns favoured by the metalworker of Dooey, Co. Donegal (Fig. 67:1a), and it made frequent appearances on the Newry latchets (Fig. 66:5), on which it became the central decoration of the orbs of the sunburst patterns. Hexafoils, in their standard form, and in multiples, appear on the reverse sides of the terminals of the Ballinderry brooch.[42] A hexafoil is also at the centre of the print on the hanging-bowl from Baginton, Warwickshire.

These are but a few examples. As a pattern, the hexafoil outlasted all others, surviving well into the Middle Ages; and it can be seen on carved whalebone from a pair of stays which are preserved in the Victoria and Albert Museum, London, and belonging to Elizabethan times.

Patterns were used by themselves or in combination with motifs. Whatever the combination, they are most likely to have been linked by scrolls, which were plain, trumpet-shaped, or broken backed; and there were almost as many variations of these as of the patterns themselves. They are best discussed by example, which will be done, each and all in their proper places, and from time to time in the following pages.

NOTES

1. T.C. Lethbridge, 'Burial of an Iron Age Warrior', *Proc. Camb. Antiq. Soc.*, XLVII (1954), p. 33, pl. VII.
2. George Macdonald, 'The Roman Fort at Mumrills', *Proc. Soc. Antiq. Scot.*, LXIII (1928–9), p. 555, Fig. 115:13.
3. For further information, see sub-heading f), infra.
4. D. Dudley, 'Excavations on Nor'nour', *Arch. Journ.*, CXXIV (1967), p. 57, Fig. 22:206.
5. A.O. Curle and J.E. Cree, 'Excavations on Traprain Law', *Proc. Soc. Antiq. Scot.*, XLIX (1914–15), p. 168, Fig. 23:5.
6. *Yorks. Arch. Journ.*, XXXIV (1939), p. 101.
7. C.F.C. Hawkes, 'Excavations at Alchester', *Antiq. Journ.*, VII (1927), p. 181, Fig. 11:3.
8. *Proc. Soc. Antiq. Scot.*, XXXV (1900-1), p. 405, pl. A/2.
9. Curle, *A Roman Frontier Post and its People*, pl. LXXXIX:20.
10. J.P. Bushe-Fox, 'Excavations on the Site of the Roman Town of Wroxeter', *Soc. Antiq. Research Report* (1912), p. 26, Fig. 10:9.
11. J.P. Bushe-Fox, 'First Report on the Excavations of the Roman Fort at Richborough', *Soc. Antiq. Research Report*, pl. XII:7.
12. Aileen Fox, 'The Legionary Fortress at Caerleon, Monmouth: Excavation in Myrtle Cottage Orchard', *Arch. Camb.*, XCV (1940), p. 130, Fig. 6.
13. C. Dickenson and P. Wenham, 'Discoveries in the Roman Cemetery on the Mount, York', *Yorks. Arch. Journ.*, XXXIX (1958), p. 317, Fig. 13:89a.
14. *Proc. Soc. Antiq. Scot.*, LVIII (1923–4), p. 251, Fig. 9:1.
15. *BM Guide to Antiquities of Roman Britain*, Fig. 11:25.
16. Rynne and O'Riordain, 'Settlement in the Sandhills at Dooey, Co. Donegal', p. 63, Fig. 8.
17. *Ulster Journ. Arch.*, IX (1861–2), p. 271, pl. I, Fig. 2; see also *Proc. Soc. Antiq.*[2], XXX (1917–18), p. 129, Fig. 14.
18. From shaft grave V, at Mycenae. Piggott, *Ancient Europe* (Edinburgh University Press, 1973), pl. XV.
19. Jacobsthal, *Early Celtic Art*, no. 19.
20. Ibid., nos. 215, 217.
21. *BM Guide to Early Iron Age Antiquities*, pl. VI:5.
22. Duignan, 'The Turoe Stone' in Duval and Hawkes, p. 201.
23. F. Haverfield in R.H. Forster and W.H. Knowles, 'Corstopitum: Report on the Excavations in 1910',

Arch Ael.[3], VII, pp. 178, 188, pl. IV:6.
24. R. Munro, *Lake Dwellings in Europe* (Cassell & Co., 1890), p. 411, Figs. 144, 145.
25. Jacobsthal[19], p. 185, no. 169.
26. The representation of paired spherical triangles on the bronze plate or belt mount from Drumashie, Dores, Inverness (Fig. 15:6) is not typical. The forms of the trumpets, which make up the main part of the decoration, suggest that this belt-plate should belong to the latter half of the first century AD.
27. *BM Guide*[21], Fig. 177.
28. P.J. Fowler, 'A Roman Barrow at Knobb's Crook, Woodlands, Dorset', *Antiq. Journ.*, XLV (1965), p. 36, Fig. 5:7.
29. H. St George Gray, 'Bronze Scabbard found at Meare, Somerset', *Antiq. Journ.*, X (1930), p. 154.
30. *BM Guide*[21], Fig. 188.
31. W. Greenwell, 'Early Iron Age Burials in Yorkshire', *Arch.*, LX (1906), p. 281, Fig. 23.
32. R.G. Collingwood, *Roman Britain* (Methuen, 1930), p. 241, Fig. 75:37.
33. C.A. Moberg, *Acta Arch.*, XXI (1950), p. 88.
34. de Navarro in *The Heritage of Early Britain*, p. 77.
35. J.M. Kemble, *Horae Ferales*, p. 192, pl. XVIII; see also *V.C.H. Lancashire*, I (1920), p. 247, Fig. 30; *BM Guide*[21], Fig. 117.
36. S. Piggott, 'Swords and Scabbards of the British Early Iron Age', *Proc. Preh. Soc.*, XVI (1950), pp. 1, 27.
37. J.D. Cowen, 'A Celtic Sword Pommel at Tullie House', *Trans. Cumb. and West. Antiq. and Arch. Soc.*, XXXVII (1937), p. 67 and plate.
38. Jan Filip, *Celtic Civilisation and its Heritage* (Czechoslovak Academy of Sciences, n.d.), p. 161, Fig. 39.
39. Jacobsthal[19], p. 202, no. 390 and pl. 266:144.
40. Behn, *Kataloge Mainz*, no. 8, no. 1080, Fig. 17.
41. As examples we may quote the altar dedicated to the god Vitiris (E.A. Wallis Budge, *A Catalogue of the Roman Antiquities in the Museum at Chesters*, p. 37); the tombstone of Gaius Sanfrius (*BM Guide*[15], p. 60, Fig. 28:VI, b); and the antefix of type Prysq 2, from Period 5 – AD 268–90 – found in the Hall at Caerleon (L. Murray Threipland, 'The Hall, Caerleon, 1964', *Arch. Camb.*, CXVIII (1970), p. 86).
42. H. O'Neil Hencken, 'Ballinderry Crannog No. 2', *Proc. Roy. Irish Acad.*, XLVII C (1942), p. 34, Fig. 13.

PART TWO: PRE- AND EARLY-INVASION STYLES

5. Introduction

We know that Tacitus did not have a very high opinion of the Britons. He found them to be disunited, divided into factions, and unable to come together in the public interest. He found them warlike, their fighting strength being in their infantry, and that some warriors took to the field in chariots, although as they fought only in parties, in time the nation was subdued. The Roman mind was not disposed favourably towards them, nor, as a last thought, towards the weather.

Of course, Tacitus was prejudiced. All the same, the picture of first-century Britain that emerges is all too familiar. At the beginning of the first century, Britain was divided into tribal areas, each with its king or prince or chief. Society was an heroic one, and in such societies heroes require fine plumes. As a result there were lavishly decorated shields, scabbards, horse-trappings, decorated chariots — many items eye-catching because of their coloured enamels — mirrors, highly ornamented and a concession to the vainglorious. The list of expensive objects is a long one. Royal warrant kept up the standard of the metal-work; and perhaps the metalworkers of one tribe vied with those of another.

Disunion made it impossible for anyone to look too far ahead, or maybe even beyond his own horizons. The Romans were lined up along the coasts of Gaul. Since the time of Augustus, Roman merchants had been increasingly penetrating Britain. Ostensibly, they came to do business; but, at this distance, quite clearly they were spies, sent to spy out the land, and to count the strength of the enemy. All this should have been clear to everyone, particularly since Caesar's foray into Britain must still have been green in the memory; yet little or nothing was done to provide defences against attack from the sea. When eventually in AD 43 the Roman forces disembarked on the beaches of Kent, at first they had some difficulty in finding someone to fight. By 47 most of the lowland zone of Britain had been overrun, and a frontier had been constructed on the Fosse Way.

The Invasion and subsequent Occupation brought about a change in the character of Britain. Once the Britons were subdued, degeneracy set in. Virtue died with expiring liberty, genius was debased and sloth took over. These changes are reflected in the quality of the metalwork, and in the cessation of metalworking on the grand scale. Shields, swords and scabbards disappeared, for they were proscribed, and thus the metalworkers were robbed of their princely patronage. The farmers were stripped of their bushels of corn, and most people were reduced to a level a little above subsistence.[1] Undoubtedly some valuables were hidden in the ground for safe-keeping, for us to discover, so we can admire the beauty of their decoration and the skills that went into making them.

The survival of the metalworkers depended on their finding new markets for new products. Clearly they would have had to quit their old workshops, and it appears that some sought refuge in the fringe areas, still unoccupied by the Romans. Perhaps most of

71

them went to Brigantia, still the largest tribal state. Brigantia lay to the north of the heavily wooded Midland plain. The Romans quickly found out that the Brigantes were formidable adversaries; so the army hesitated to invade. Instead, the Romans consolidated their hold on the territory south of an imaginary line joining York to Derby to Chester. But, with the construction of the Fosse Way Limes in 47–8,[2] the Iceni in Norfolk became disillusioned at what they regarded as encirclement,[3] so their disarmament followed. Also in 47 Caratacus was able to rouse the Silures and the Ordovices in Wales to aggression, but the trouble was put down by Ostorius, who, in 48, advanced towards the Cheshire Gap. Here he divided the tribes of Wales and the Marches from those of the Pennines, and by so doing he cut the source of aggression of the year before. Dissidents of the times must have found the Cheshire Gap to be a very useful highway between Wales and Brigantia.

In the territory of the Iceni, the process of Romanisation was cut short by a further uprising in AD 60, when, for a very short time, 'the history of Icenia was also the history of Britain'. Not less than 70,000 people were put to the sword without distinction. A number of important metal hoards were buried about this time, and generally in eastern England.

Such historical events may appear dry as dust to most readers; but the metalworkers had to live through these times, and to a greater or a lesser extent their lives must have been affected by these events. Even marital quarrels exacted their toll. The quarrels of Cartimandua, Queen of the Brigantes, with her consort Venutius, brought forth Roman intervention. The tribesmen rallied to Venutius, but, with Roman help, Cartimandua regained control. Venutius kept to the background for a decade, during which time a town, Isurium Brigantium, sprang up at Aldborough. The link with Rome allowed the natives to form settlements near the forts, thus permitting the sale of native-made wares to the world beyond. The result was that a strong commercial link between the native craftsmen and the Roman world was established, whilst natives living out in isolated farmsteads were also seeking Celtic metalwork.

So, after the initial troubles, craftsmen were once more in a position to manufacture and to sell. The Roman world was at their feet, and they had the home market as well. That they seized the opportunities offered is beyond question. Princely favours had now been exchanged for a stand in the market-place. The public were the buyers, and production was going up. Even in the far-away Votadinian oppidum of Traprain Law expansion was afoot. Everything seemed to be set fair for the future. Yet this was only a preparation for the disaster of AD 74.

What follows will cover the styles that were current in the early part of the first century AD, together with those of the early years of the Occupation period.

NOTES

1. Hawkes, 'Britons, Romans and Saxons round Salisbury', p. 79.
2. *Journ. Roman Studies*, XIV (1924), p. 256.

3. T. Davies Pryce, 'The Roman Occupation of Britain: its early Phase', *Antiq. Journ.*, XVIII (1938), p. 40, footnote 1.

6. The Galloway Style

Celtic Art is not without its variations, notably of interpretation. Variations in the interpretation of motifs and patterns are often on a regional basis; and some, no doubt, were occasioned by the whims of the local metalworkers. Because the country was divided up into tribal areas, this situation too could have had some bearing on these differences. For perhaps these reasons, the products of the Galloway metalworkers must be seen to have been in a class of their own, both technically and imaginatively. The decorative motifs and patterns used are in every sense in the best traditions of Celtic art, yet they differ from all other British-made products. For this reason, we can speak of a very individual style — the Galloway style.

The area in which these products originated lies to the north of the Solway, taking in modern Dumfries and Kirkcudbright. Seen from south of the Solway Firth, this region appears to have an air of remoteness. Yet, from the west coast of Galloway, Ireland is clearly visible. To guard against trouble from that quarter, in AD 82 Agricola 'lined this coast with a body of troops, not so much from any apprehension of danger, as with a view to future prospects'.[1] From the Irish side of the channel, Galloway also held out prospects 'by the means of [Irish] merchants resorting thither for the sake of commerce, the harbours and approaches to the coast are well known'.[2] In addition, the Irish had friends and relations in Galloway, and these were called the Cruithin, who were cousins of the Cruithin of Ulster. No small wonder Agricola was cautious. These comings and goings and relationships make up the background to some of the most spectacular finds ever made in Britain.

It used to be thought[3] that La Tène chieftains had settled in Galloway by the end of the third century BC, though it was not known whence they came, though Yorkshire was suggested. But Yorkshire has yielded insufficient comparative evidence to make this belief credible. Galloway was rich in natural resources, enough to encourage the establishment of workshops here; there are copper ores in the foothills at the back of Wigton Bay. The Cruithin were competent metalworkers, as we know, and they were in a position to exploit these resources. In Ireland, their first cousins, the Soghain, were the holders of the Celtic Art franchise. So the production in Galloway of items such as the Torrs pony-cap and the Lochar Moss beaded collar are in no way surprising.

Another product is the bronze collar which was found on the banks of the Eden, near Stichill, Roxburghshire[4] (Fig. 19). Collars succeeded torcs; and whilst this specimen was found in an area remote from Galloway, nevertheless it is made in the Galloway style. Like some other collars, it is hinged at the back. It was found in association with a pair of massive armlets, suggesting the possibility of its having been purloined by a raiding Caledonian.

Fig. 19: Decorated Bronze Collar, Stichill, Roxburghshire. (1/2)

The collar is said to be late because of the hinge at the back. However, the armlet from Plunton Castle (Fig. 15:2) is similarly hinged, so that Roman influence need not be held responsible for the presence of this hinge here. This is not to deny that hinges can occur later.[5] Another similarity between collar and armlet is in the occurrence of swash Ns amongst the decoration in both cases. These swash Ns are contained within small panels, supposedly riveted on. In the case of the Plunton armlet, the pattern resulted from good die-work, the patterns on both panels having been produced from the same die. In the case of the collar, this was not possible, because of the tapering form of the collar itself, necessitating a certain curtailment of the pattern on one side. There has been skilled setting out of the decoration, which is in the repoussé style. One immediately thinks of the repoussé decoration on the Torrs pony-cap, and the admitted similarities between the two must be noted down. The pattern on the left hand terminal of the collar resembles the earliest continental forms incorporating the lotus-bud motif, whilst others are reminiscent of those on the largest Polden Hill mount. The right-hand motif is in most respects like another on the pony-cap[6] (Fig. 13:1). The only essential difference is in the matter of date; for the Torrs pony-cap was fashioned at a time when the boss style was emerging, whereas, in the case of the Stichill collar, that period of development had passed. The Torrs pony-cap is thus earlier than is the Stichill collar.

But, what of the Torrs decoration? The lotus-bud motifs have more than a passing resemblance to others on the Turoe stone. But the Torrs motifs are not what at first sight they appear to be: for, instead of being simple, they are complex, being made up of two motifs, the cable motif and the lotus-bud motif; but here the identity of both is being lost, with the result that the single representation is a little like a Brentford palmette.[7] The mixing of motifs was noted amongst the decoration on the Aesica brooch; and what one is left with is a bastard form, which is neither one motif nor the other. All in all, the Torrs pony-cap must be looked upon as a link between the decoration on the Stichill collar and that on the Turoe stone. Here one should pause to consider the implications of the Turoe decoration having offshoots in Galloway and Roxburghshire.

This situation is perhaps significant; but significant for what? Not so much for the art, as for the distribution of a style. Not so much for the distribution of a style as for the nature of its origin. Origins are not always easy to trace. In the case of the Turoe stone, and on a wholly immovable object, there is a repository of motifs — cable motifs, lotus-bud motifs, triskeles and so on, and there are even the makings of an emergent boss style,[8] which here is more akin to that on the Llyn Cerrig Bach plaque and Llandyssul collar. It is not strictly a northern style, as defined by Leeds.

The Torrs lotus-bud motifs have that quality that was first noted at Brunn-am-Steinfeld. However, another and a more classical style was creeping into Britain, and this can be seen on the Polden Hill mount (Fig. 23:1). It is also well represented on the Aesica brooch, the Housesteads silver brooch and the Newstead buckle. Its appearance on the Stichill collar may have heralded a slight cultural change, in that Irish influence was lessening. But this impression could hardly be correct, because now Irish influence had never been stronger,[9] and was making itself felt in the Galloway workshops, as a second look at the Stichill collar will convince: because two manufacturing styles are represented. The second style differs considerably from the first: for the first is made up of heavily embossed designs, whereas the second style is more effete, certainly less robust. Representative patterns can be found on both sides of the hinge, and there are more of them running down both sides of the collar. All of it is tooled in accordance with Irish practice. The nearest parallel is to be found on the Cork horns[10] (Fig. 13:5), as Stevenson has already pointed out. The decoration on both of these objects is remarkably alike, with the Cork design perhaps being a little more mature, and therefore more developed. The bosses are larger, more like those on the terminals of the collar, whereas the design of the swash Ns is made up of trumpets and lotus-buds. Though trumpets occur on the Stichill collar, they are nevertheless still in embryo form, in contradistinction to those on the Cork horns, on which they appear in developed style. Further, the Cork decoration has been applied by the same method used for that on the Bann disc[11] — that is, by casting. If the Cork horns appear to be in a very remote area, then let it be remembered that a small colony of the Soghain once lived here. It is possible they brought these horns with them when they left the 'old plain of the Soghain'.

Irish decorative work is known to be in two styles: there is the robust embossed or repoussé style of the so-called sacrificial dishes (Fig. 5:2); and there is the somewhat effete and often spidery style of decoration present on the Cork horns, the Petrie Crown and the Bann disc. Robust or spidery, Irish decoration is often given over to exaggeration, though complete artistic control was maintained.

It is unusual for two styles to be present on the one object, and in this respect the Stichill collar might well be unique. With one style indubitably Irish and the other at home in Galloway, the disharmony of the two styles can best be explained by the certain fact that one was added at a later date. Taken as a whole, the decoration will be found to have anomalous elements, and until the above fact is accepted, the anomaly will remain. The Cork horns type of decoration was developed out of the Battersea shield style, in which the central boss is decorated with thin, spidery designs in repoussé style,[12] a style that includes a simplified version of the palmette. Small bosses proliferate, and there are the beginnings of the elongated trumpet style. On the upper and the lower roundels of the shield, the swelled ends of the scrolls are executed in the same style as that seen on the Torrs pony-cap, and suggested in the first place by decoration on the Turoe stone. All these small features are early in Britain, and from the suggested date of the Irish style, which could be as late as the early second century AD, it will be appreciated that there was adequate time for development. The Irish style is dated by the incorporation into the pattern on the Petrie Crown of an idea developed by the Brigantes. This is seen in the form of two heads (Fig. 35:1), borrowed from dragonesque brooches, of a form current in the early second century.[13] Since there is a unity of style amongst the decoration on the Cork horns, the Petrie Crown and the Bann disc (on the last named, these heads are breaking up), it would appear that this spidery form of decoration, which appears also on the Stichill collar, must, in the last case, have been carried out by somebody working at a later date than he who applied the robust Galloway style decoration.

Both styles must have been reserved for exceptional objects, objects of great ostentation and the search for parallels produces none. The early first century is suggested for the bold Galloway style: but in Galloway itself we are at the end of the road, since, for a continuation of this repoussé style we have to go to Lough-na-Shade, Co. Armagh,[14] where a trumpet mouth was elaborately decorated in a modified version of the Galloway style. Part of this decoration is shown in Fig. 13:3, from which it will be noted that there has been a toning down of the style. The lotus-bud motif is weak and thin by comparison with those on the Torrs pony-cap, whilst the tendril has been given a twirl that must herald the forthcoming snail-shell bosses of the Broighter torc. Also there is a link with the broken back triskele of the Tal-y-llyn form (Fig. 22:3), which in its turn has certain resemblances to the Galloway style, here seen in the treatment of the swelled ends of the scrolls.

The Lough-na-Shade trumpet comes from the territory of the Soghain, who were very busy at Lough Crew sketching designs incorporating broken-back scrolls, and all done with the aid of a compass. These drawings (Fig. 34) were recorded for us by being applied to bone 'trial pieces'. They must have served as patterns for decoration on metalwork; yet none of this survives. But for their being on bone, there would have been no record of these patterns. Their occurrence should serve as a salutary warning to anyone wishing to make too much out of the absence, or near absence, of good metalwork. The broken back scroll on the plate of the Lochar Moss beaded torc[15] is another link with Irish art. These scrolls (Fig. 38) were fabricated separately, being applied afterwards and kept in position with the help of domed rivets, which look like bosses, and doubtless were intended to. The idea was a labour-saving one, and in some ways is indicative of a very modern approach. The mode of applying decoration as in the case of the Bann disc is also a labour-saving idea. This situation must indicate that the metalworkers were now searching for labour-saving ideas, and were adopting a very modern approach to the problem.

It is now evident that the flamboyant nature of the decoration on the less than practical Irish products is also a characteristic of the Galloway-style objects. Perhaps the one British (as opposed to Galloway) product that comes nearest in style is the horned helmet from the Thames at Waterloo Bridge.[16] The Galloway style was going through a process of alteration, as was Irish decoration at that time; but it managed to persist into the second century, if the western fringe area crannog evidence means anything at all. Agricola's stationing of troops along the west coast of Galloway must seriously have interfered with communications between that area and Ireland. An object made at about this time is the Balmaclellan mirror mount[17] (Fig. 5:3), on which elongated Irish type trumpets appear, a little like those seen on the 'sacrificial' dishes; but other influences were at work, and by this time the craftsmen were beginning to look eastwards. On the mirror mount there are rosettes and a beaded edge. Rosettes occur at Stanwick, and beaded edges are typical edgings for Brigantian disc-brooches of the second half of the first century. It is clear that Agricola's men were performing their job well.

The Galloway-style story ends with the punitive measures undertaken by the Romans in that region in AD 122.

NOTES

1. Tacitus, *Agricola*, XXIV.

2. Ibid., XXIV.

3. Sheppard Frere, *Britannia* (Sphere Books, 1974), p. 29.

4. D. Wilson, *Archaeological and Prehistoric Annals of Scotland* (Sutherland & Knox, 1851), p. 451 and illustration; Anderson, *Scotland in Pagan Times* (David Douglas, 1883), pp. 135–7, Fig. 112; *Royal Commission on Ancient and Historical Monuments, Roxburghshire*, p. 22; *Proc. Soc. Antiq. Scot.*, XXXVIII (1903–4), pp. 462–6; Atkinson and Piggott, 'The Torrs Chamfrein', p. 232, pl. LXXXV:b; Fox, *Pattern and Purpose* (National Museum of Wales, 1958), p. 107, pl. 62:b; R.B.K. Stevenson, 'Metalwork and other Objects in Scotland' in A.L.F. Rivet (ed.), *The Iron Age in North Britain*, p. 32, pl. 6; Megaw, *Art of the European Iron Age*, no. 298.

5. Hinges are known from Roman metalwork.

6. Atkinson and Piggott[4], Pl. LXXXI.

7. As seen on the Brentford horn cap. This work Fig. 29:2 and *Arch.*, LXIX (1920), p. 22, Fig. 22.

8. Duignan, 'The Turoe Stone' in Duval and Hawkes, Fig. 4.

9. This had been noted before by Stevenson[4], p. 32, and Megaw[4], p. 172.

10. M.J. O'Kelly, 'The Cork Horns, the Petrie Crown and the Bann Disc', *Journ. Cork Hist. and Arch. Soc.*, LXVI (1961), pl. IV.

11. E.M. Jope, 'The decorative cast Bronze Disc from the River Bann near Coleraine', *Ulster Journ. Arch.*, XX (1957), p. 95 and illustrations.

12. Best illustration in Fox[4], pls. 16 and 17.

13. H.E. Kilbride-Jones, 'The Excavation of a Native Settlement at Milking Gap, High Shield, Northumberland', *Arch. Ael.*[4], XV (1938), p. 342.

14. F. Henry, *Irish Art* (Methuen, 1947), pl. 7a.

15. A. Way, 'Notice of a bronze Beaded Collar found at Lochar Moss, Dumfriesshire', *Arch.*, XXXIV (1852), p. 83, pl. XI.

16. Fox[4], p. 50, Fig. 36a.

17. Wilson[4], p. 228, Fig. 147; Anderson[4], p. 127, Figs. 103 and 104.

7. The Elmswell-Broighter Style

A very remarkable panel, intended for covering an iron box or casket, was found in a native settlement of the Parisii at Elmswell, which is about two miles west of Driffield, in east Yorkshire. This settlement had been occupied from at least Flavian times, up till the end of the Roman Occupation.[1] At a later period, in Saxon times, it was occupied again. Unfortunately, the panel (Fig. 20:1) was not stratified.

This panel is covered with repoussé decoration, which has been hammered up from the back. The technique immediately brings to mind the Galloway style, and the same technique was used on objects like the Broighter, Cairnsmuir and Snettisham torcs, the Lough-na-Shade trumpet mouth and the Deskford boar's head. But the Elmswell panel is also remarkable for having a top strip carried out in orange-red champlevé enamel. This could indicate Belgic-Romanising influence, which Corder and Hawkes recognise as a genuine cultural infusion. However, it must be remembered that, in the north, there is no evidence for champlevé enamelling before the middle of the first century, at least not in this form. The same authors believe that the design on this strip is a rendering of the Roman vine-scroll, as seen on early Samian ware; and such an occurrence could indicate a pre-Flavian date for the panel. The design is not all Roman: included in it, at the extreme ends of the panel, are two small lotus-bud motifs, here complete with tendrils. It is quite clear that this enamelled strip does not run in naturally with the composition below, which is in repoussé style. But Corder and Hawkes prefer to regard this as a sign of early immaturity, and therefore they date the panel to between AD 50 and 70.

Bold repoussé patterns such as this were beaten out on a hardwood mould. The wood partly absorbed the hammer blows, necessary if the metal was not to be breached by the tool. The sweeping curves and the general 'roundness' of the pattern are the necessary confirmation. Overall, the pattern is repetitive, and basically it consists of two motifs: the cable motif and the double trumpet formed of two conjoined continental trumpet motifs. There is some relationship to the boss style here, though it is in effect divorced from it; also it stands apart from the Galloway style with which it might have been associated, in view of the likeness of the decorative technique. But, where bosses should have been used, there are rosettes instead. These rosettes are of the berried kind, and, as such, they make only a limited appearance in British art. They occur on pottery from Hunsbury;[2] on a very native-looking weaving comb from Newstead;[3] and on bridle-bits from Ulceby, Lincolnshire,[4] and Ringstead, Norfolk.[5] The Ulceby specimen is shown in Fig. 21:1. Berried rosettes can also be seen on the fan-tail brooch from Tre'r Ceiri,[6] and on a disc-brooch (Fig. 20A) in the Newcastle Museum.[7] But though appearances of the berried rosette are somewhat limited in Britain, these rosettes are of great antiquity on the Continent. They make up the central decoration on a corselet or girdle mount from

78

Bergamo,[8] and again on the gold framework of a Scythian leather cap from Ositujazka.[9] With these continental and Brigantian occurrences it is possible that the berried rosette was brought to the north-east by new arrivals. There was always room in British art for new patterns, since the metalworkers' taste was voracious of alien motifs or patterns, which they appear to have preferred to their own indigenous forms.

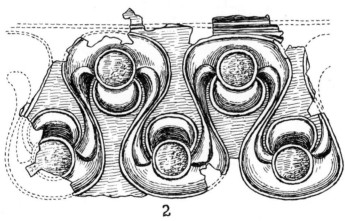

Fig. 20: Bronze Panels with Repoussé Decoration 1. Elmswell, east Yorkshire. 2. Great Tower Street, London. (1/2)

In point of time, the berried rosette must have given way to the form seen on the Balmaclellan mirror mount (Fig. 5:3) which are themselves similar to those occurring at Stanwick. This suggests a natural sequence of events, since the Balmaclellan mirror mount is clearly later, as indeed is the rosette on the Nijmegen mirror handle[10] (Fig. 21:2). Also on this mirror handle, we are introduced to another feature, the raised quadrant. The central rosette is flanked by two of these quadrants, here cast, since this is a cast handle. But raised quadrants are common in repoussé work, and they lend themselves very well to that style: they appear on the Elmswell panel, and again on another from Great Tower Street, London[11] (Fig. 20:2). They occur on the Ringstead bridle-bits cited above. To some extent, quadrants and berried rosettes go together, and they may be regarded as two elements of the style which forms the subject of this chapter.

Fig. 20A: No locality (Newcastle Museum 1956: 160A). (9/4)

However, it is almost a dictum in Celtic art circles that once two elements, such as these, appear together on a single object, then one will be dropped in preference for the other. This is the case with the Deskford boar's head, on which raised quadrants are used to represent eyelids, but rosettes are missing. Quadrants are emergent amongst the decoration on the helmet of unknown provenance, but thought to be from northern England,[12] and which is now in the British Museum. Again, they are found amongst the decoration on the massive armlets, which itself is clearly derived from the repoussé style of the first century, and of which good examples are shown in Fig. 9:2 and 3. However, raised quadrants do not appear on the Broighter torc, but make tentative appearances on such items as the so-called sacrificial dish from Monasterevan, Co. Kildare (Fig. 33:2) and the disc shown in Fig. 33:1, on which they are used as curved links in the overall pattern. The Monasterevan disc decoration is highly devolved; yet even here the craftsman could do better, as he did when he executed just one trumpet form. The 'twirl' given to the boss at the top surely presages the snail-shell form of boss which is common on the Broighter torc. The amount of interplay in the decoration appearing on objects found on both sides of the North Channel is very marked, and it can be put down to a common tradition. The raised quadrant which occupies a central position on the Balmaclellan mirror mount is

Fig. 21: 1. Ring of Bridle-bit, Ulceby, Lincs. 2. Handle and Mount from Decorated Bronze Mirror, Nijmegen, Holland. (3/4)

similar to those on the Monasterevan disc and the Deskford boar's head. The extension to include the Broighter torc is perhaps more academic, but it is based not only on the snail-shell boss being a development out of 'twirls' seen on the disc, but also on the fact that the representative gold torc terminal decoration seen on the Ulceby bridle-bit (Fig. 21:1) is also associated with berried rosettes.

Another characteristic of the Elmswell-Broighter style is the crowding in of as much decoration as possible on the amount of space available. Rosettes, quadrants, devolved cable motifs and poorly represented trumpets elbow each other for position, but the process is orderly. The Elmswell, Great Tower Street panels as well as the Broighter torc are typical of what was achieved. All are exceedingly pleasant to look at. The same desire to cover every available inch of space is a dominant feature of the Turoe stone, on which the cable motif is prominent. The treatment of the motif here is very similar to its treatment on the objects named. The Elmswell panel also has the distinction of having double trumpets amongst its decoration. One pair is stretched out at the centre bottom of the panel: others occur to right and left of centre, near the top. Elsewhere there are representations which seem to have been inspired by the same idea. The relationship of the Great Tower Street panel to the one from Elmswell is based on the fact that the entire decoration is taken from the central section of the design on the Elmswell piece. By placing both drawings, one above the other, as seen in Fig. 20, this point emerges clearly. It will also be seen that the berried rosettes have given way to filled non-bosses.

So, if we talk here of an Elmswell-Broighter style, we do so because of the strong links that are evident in the decoration of the pieces named. Almost everywhere, the cable motif is the basis of the patterns, as it is on the Snettisham gold torcs. A peculiarity of the style is that the cable motifs utilised are always associated with features that are plainly unusual: on the Broighter torc with snail-shell shaped bosses; on the Snettisham torc with a type of rosette that can be paralleled on the Aylesford bucket; on the Elmswell panel with berried rosettes; and, in the case of the Great Tower Street panel, the rosettes have been replaced by filled non-bosses. Each and all are different, one from the other, and there is an air of searching and experimentation. Normally, cable motifs are plain, with their thin ends partly wrapped around a plain boss. The fanciful interpretation of the motif is typical of the Elmswell-Broighter style, and the most fanciful of all exercises is represented by the decoration on the Broighter torc. Another characteristic, noted everywhere, is the very strong relief that has been given to decoration as a whole; so much so that it is a miracle that the metal did not crack or fracture more than it has done. The metal of the Elmswell and Great Tower Street panels has been tested to the full. Gold can stand up better to this sort of treatment, which might suggest that the metalworkers were in fact goldsmiths. Gold has an ability to stretch which bronze lacks. So that all these fanciful effects, adding up to a separate style, may have been thought up as being suitable for gold as a medium. The style is undoubtedly early, so that Corder's and Hawke's date for the Elmswell panel might be a little late.

Two more objects call for inclusion within the same style. The first is the helmet which is thought to have come from northern England, and now housed in the British Museum. The pattern on the neck guard is very similar to that on both the Elmswell and Great Tower Street panels, but with this difference: the character of the cable motifs present has altered. Here these motifs are acquiring hollow centres, like those on the Turoe stone.[13] Some might prefer to call these hollows enclosed voids. These pelta-shaped voids are beginning to split open the cable motifs in this helmet decoration, but the resulting effect is not unattractive. To give added effect, there is a heightened ridge, which helps to throw the work into strong relief. The criss-cross pattern on the studs is also found on the tankard handle from Seven Sisters, Glamorgan[14] and again it can be seen among the patterns recorded on the Lough Crew trial-pieces (Fig. 34:12). Here is another oddity to add to those mentioned above; and, apart from the Irish connection, there is the interesting point that, in the case of the Seven Sisters tankard handle, it forms part of an emergent boss-and-petal style that was generally current not later than the first half of the first century AD.

Up till this point, all decoration in the Elmswell-Broighter style has been carried out in the repoussé style. But the second object mentioned above, and now coming up for consideration, is cast. So we have a casting, as opposed to decorated sheet-work. The enamelled mirror mount (Fig. 21:2) from Nijmegen[15] is an impressive piece, with nicely balanced decoration; and that decoration is nothing more than a modified version of what has been seen on the Elmswell and Great Tower Street panels. The sole difference is one of technique. Here are the last and lasting remnants of the cable motif (probably everyone was getting tired of it by this time) associated with filled non-studs (or grey paste filled bosses, as some might prefer to call them) which are a simple substitute for the rosettes. This mirror is generally regarded as being the last of the British mirrors, and it is also the largest. It was found in the Roman legionary fortress of Numaga, in a grave which also contained a glass vessel with M-handles, and Gallo-Belgic ware of the first century AD. The glass vessel is of first/early-second-century-AD type.

In retrospect, it can be said of the Elmswell-Broighter style that it was sophisticated; but, like so many sophisticated styles, its life was not unduly long. Perhaps it fell victim to the political or military fortunes of the times. Its repoussé-style decoration shows that it was early; and with its final occurrence on the Nijmegen mirror, it can be said to have had a life-span of almost a century. The decoration on the Broighter torc must represent its real beginnings, which puts the genesis of the style into the pre-Occupation period.

NOTES

1. P. Corder and C.F.C. Hawkes, 'A Panel of Celtic Ornament from Elmswell, east Yorkshire', *Antiq. Journ.*, XX (1940), p. 338.

2. Fox, *Pattern and Purpose*, pl. 77c.

3. Curle, *A Roman Frontier Post and its People*, pl. LXVIII, no. 4. Found in Pit LIX, associated with a cup of Form D. 27, and portions of Samian Form D. 29, all of Flavian date.

4. Fox[2], pl. 24.

5. T.G.E. Powell, *The Celts* (Thames & Hudson, 1958), pls. 37, 38.

6. Fox[2], p. 107, Fig. 70. Brooches of the fan-tail group are manifestly an adaptation of the Roman thistle form: as a type they died out about the middle of the first century.

7. Macgregor, *Early Celtic Art in North Britain*, I, Fig. 5:7.

8. Jacobsthal, *Early Celtic Art*, p. 178, no. 134.

9. Ibid., pl. 293a.

10. G.C. Dunning, 'An Engraved Bronze Mirror from Nijmegen, Holland', *Arch. Journ.*, LXXXV (1930), p. 69, Fig. 1.

11. Corder and Hawkes[1], pl. LIII:a.

12. Fox[2], pl. 62:c; J. Brailsford, *Early Celtic Masterpieces from Britain* (British Museum), Figs. 53, 54.

13. Duignan, 'The Turoe Stone' in Duval and Hawkes, p. 207, Fig. 6:c and e.

14. Fox[2], pl. 66:a.

15. Megaw, *Art of the European Iron Age*, p. 156. Megaw compares the Nijmegen decoration with that on the Santon mounts, but he has missed the significance of the raised quadrants enclosing the rosette. This same handle decoration was copied in Ireland, on a small brooch found at Knowth; and there is a further reflection of it on the Amerden scabbard. Quadrants of the same form also occur on a first-century tankard handle from Stoke Abbot, Dorset.

8. The Aesica Style

It is now many years since Collingwood[1] wrote that 'in the development of any type, once established, decadence is the general law of art'. Nothing could illustrate this dictum better than the Aesica style. The Aesica style is intensely vigorous at its beginnings, with full development attained at the time the Aesica brooch was made; but thereafter it went into a steady decline.

The Aesica brooch (Fig. 4) has already been partly discussed. Its repoussé decoration incorporates two underlying motifs: the lotus-bud and the continental trumpet motif. The former appears variously, either singly, paired (though in opposition) or conjoined to the trumpet motif. The conjoining of two different motifs is not new: there is a combination of a lotus-bud and cable motif on the Torrs pony-cap. But the conjoining of a lotus-bud and trumpet motif is something that is new; but still it is a combination that was attempted in the decoration on the reverse sides of the terminals of the gold torcs from Cairnsmuir and Snettisham (Fig. 6) and appearing in isolation, hidden at the back, as though its message was a secret that only the wearer knew about. All representations of these motifs have been most carefully carried out, in somewhat stronger contrast to the random though carefully tooled decoration on the obverse sides, where thoughts have gone astray, and the goldsmiths have left the purity of the mainstream for the slightly muddy eddies of the pools. This was an art world full of abstraction. Other curious combinations are the conjoined cable and continental trumpet motifs seen on the Balmaclellan mirror mount (Fig. 5:3). It seems that at this period, which is about the middle of the first century AD, it was the recognised thing to do; so we have here a general movement to conjoin motifs with different backgrounds, and presumably different meanings; and like everything else examined so far, this general movement could have started at Turoe.[2] The middle-first-century date for Britain is confirmed by the decoration on the Balmaclellan mount, on which are rosettes which can be paralleled at Stanwick. The same period would have seen the manufacture of the Aesica brooch, in spite of the circumstances of its discovery.

The decoration looks to be in the repoussé style, which in itself is an important fact, for in this there is almost certainly a connection with the Galloway style, or with the Elmswell-Broighter style, but only in the matter of the techniques employed. Even Stanwick must not be forgotten, since the same repoussé style is noted there too (Fig. 5:4 and 5). A further link with the Elmswell-Broighter style is evident in the little snail-shell bosses that occur on the Cairnsmuir and Snettisham torcs; whilst the compass-produced background to the decoration on the Broighter gold torc is repeated on the Snettisham gold terminal discussed in the last chapter. So there could be some argument here for a common artistic background, were not the style of another Snettisham torc decoration, that on the terminal illustrated in Fig. 6:2, seen to be completely at variance with that of the first mentioned torc: for Fig. 6:2 can be put alongside the Cairnsmuir torc, in view of the similarity of the snail-shell bosses, but not alongside Fig. 16:7. It therefore looks as though all the styles being discussed here are offshoots from a single trunk, which had its roots in the old plain of the Soghain in Co. Galway, Ireland; and the measure of the existence of these styles in Britain was due entirely to 'the (Irish) merchants resorting thither for the sake of commerce', as Tacitus has told us; and the sum total of this movement must be the presence of the Cruithin in Galloway, from amongst whom came competent metalworkers to give us the styles commented upon.

In Brigantia, these styles met increasing provincial Roman influences; and one metalworker who partly succumbed was the maker of the Elmswell panel. Another seen to have been suitably impressed was the maker of the Aesica brooch, which, basically, is a fan-tail form, derived from the Roman 'thistle' brooch, which itself was descended from Gallic and German types of the first century AD. The date of the Aesica brooch can be narrowed down to the period before AD 74, from purely historical evidence. The style of the conjoined lotus-bud and continental trumpet motifs at the top of the brooch decoration is clearly later than that of the gold torcs; for, whilst in the case of the Cairnsmuir and Snettisham torcs there are simple unions of lotus-bud and continental trumpet motifs, in the case of the Aesica brooch there is what amounts to a three-way union, in that the same combination has been placed above and on an underlying bit of repoussé work that looks suspiciously like a cable motif. There cannot be much of a gap here, so that the torcs may be said to belong to the first half of the first century, whilst the Aesica style is a mid-first-century style, and as such its history and that of the Elmswell-Broighter style must have run parallel for a time.

The little embossed disc-brooches, which Leeds[3] thought should belong to the 'Revival' period, must here be included within the Aesica style. Common to all are beaded edges, or pellet borders as some prefer. There is a significant list of discoveries on Roman sites, but this does not mean that they are Roman. Perhaps the disc-brooch form was Roman inspired, like the form of the Aesica brooch. However, the decoration is patently Celtic, with a dependence on triskeles and lotus-bud motifs. The Balmaclellan mirror mount also has beaded edges; and something approaching the same thing can be seen on the Aesica brooch. In the course of decorating small objects like disc-brooches in the repoussé style, recourse must be made to the use of thin sheet metal; otherwise it would not be possible to emboss such small details of decoration. In the case of these disc-brooches, the metal is excessively thin, necessitating its being mounted after decoration on to another disc of thicker metal, to which pin and catch are attached, in order to give strength and stability to the brooch.

A glance at Fig. 4 is sufficient to show that the style of decoration is in every way in the same tradition as that displayed on the Aesica brooch; so it is possible to speak of an Aesica style. The brooches illustrated support very well Collingwood's law regarding decadence. The best of them is the specimen from Silchester[4] (Fig. 4:2). It will be remembered that, on the Aesica brooch, the lotus-bud motifs have attached tendrils, to give them movement; and that these tendrils went beyond the metal's edge. On the Silchester brooch the tendrils have been turned in upon themselves, and already it is clear what is likely to happen. The process is made clear in the case of the disc-brooch (Fig. 4:3) from the Victoria Cave, Settle, on which the whole design is already starting to break up, and some sections have become detached. This process goes on, until the final break-up, when all vestiges of the lotus-bud are lost, and it is the tendrils that have taken over (Fig. 4:4),[5] this drift taking us fast towards the later spirals and scrolls. Scrolls and spirals are seen on some early New Forest wares,[6] which generally belong to the period *c*. AD 70–100. This could also be the period of the disc-brooches.

The triquetral layout of the decoration on all these brooches is clear: the triskele is the basis of all patterns, and any disc-brooches not having decoration set out in this manner cannot belong to the Aesica style. Though short-lived, as a style, the whole conception of it is grandly Celtic. The style is at home in the north, in spite of the brooch distribution; for two brooches which appear to be the earliest, for example, those from Silchester and Richborough[7] (Fig. 4:2 and 7) have the most southern find-spots. Three others, of which one is illustrated, Fig. 4:4, came from Brough, known to be a centre of second-century metalworking. The overall size of the brooch has decreased, a process carried a stage further at Aesica (Fig. 4:5). Parallel with the reduction in size went the simplification of the overall pattern, until little beyond the basic triskele was left, as for instance here at Aesica. Another apparently similar, and represented only by a fragment, came from Arbeia (South Shields),[8] which also yielded an earlier form, comparable with Fig. 4:3, which is the brooch from the Victoria Cave, Settle.[9] Another disc-brooch comes from Corbridge.[10] Corbridge, Aesica, Arbeia, Brough, Settle — the type was undoubtedly Brigantian, like the Aesica style itself. The translation of two of these disc-brooches to Silchester and Richborough may have resulted from their being on the persons of military personnel. The maximum distance travelled by a Brigantian product was by the circular dress-fastener, which was found at Masada: so that too much should not be read into the occurrences of disc-brooches at Silchester and Richborough.

Even if Collingwood's dictum has been established by what happened to the Aesica style, nevertheless, whilst it lasted, it is the only post-Invasion style that showed power and imagination. The method of production was an involved one, exacting and fiddly by turn, and decidedly man-hours consuming. These brooches were probably superseded by the umbonate disc-brooch, considered more attractive by many because of its coloured enamels.

NOTES

1. Collingwood, 'Romano-Celtic Art in Northumbria', p. 40.

2. Duignan, 'The Turoe Stone' in Duval and Hawkes, p. 206, Fig. 3:d.

3. E.T. Leeds, *Celtic Ornament* (Oxford, 1933), p. 139.

4. J. Romilly Allen, *Celtic Art in Pagan and Christian Times* (Methuen, n.d.), p. 108 and figure.

5. On the specimen from Brough, Cumbria. *Proc. Soc. Antiq.*[2], III (1864–7), p. 256.

6. In burial group 109 at Baldock, Herts. See W.P. Westall, 'A Romano-British Cemetery at Baldock, Herts', *Arch. Journ.*, LXXXVIII (1931), p. 276, Fig. 5:4844.

7. J.P. Bushe-Fox, 'Fourth Report on the Excavations at Richborough', *Society of Antiquaries Research Report*, XVI, p. 139, pl. XLV:170.

8. Information and drawings kindly supplied by Lindsay Allason-Jones.

9. A. King, 'Romano-British Metalwork from the Settle District of west Yorkshire', *Yorks. Arch. Journ.*, XLII (1971), p. 412, pl. Ia.

10. *Arch. Ael.*[3], V (1909), p. 406, Fig. 22.

9. Broken-back Scrolls and Triskeles

From time to time Celtic metalworkers abandoned tradition and then they produced something that is at once interesting because it is odd. One such production is the open-work bronze disc[1] (Fig. 17:6) which is now housed in the Ashmolean Museum at Oxford, but which is nevertheless without locality. From whatever angle this piece is approached, the curves will be seen to be broken backed. This is a generally accepted term used for describing certain scrolls which differ from the general run, in that suddenly there is a pause in direction, and thereafter continuation is on a different course. The best-known examples of the form, and oft quoted, are those which occur on the beaded torc from the Lochar Moss (Fig. 38).

However, there is no connection whatever between the Lochar Moss torc's broken-back scrolls and those on the present unlocalised disc. The design at Lochar Moss also included bosses, albeit small ones, which are, in effect, the heads of rivets, there to keep this openwork pattern attached to the plate. The unlocalised disc has no bosses, but once it had ornamental buttons (the rivet holes for attaching them can still be seen). These buttons were enclosed by penannular mouldings, which are in fact badly devolved cable motifs paired; and more of their kind can be seen on the trace hook from Polden Hill[2] (Fig. 28:5). A mid-first-century date is suggested for these pieces.

Craftsmen treated their broken-back scrolls with respect and not a little cunning. The man who doodled on the Loughcrew trial pieces (Fig. 34) was intent on seeing how many variations on this theme he could achieve. The man who made the Ashmolean disc wished his work to resemble the branches of a tree — a sort of naturalistic abstraction that modern artists admire. Was this a one-off effort, or had it been copied? There are no obvious descendants, so it is worth looking round to see if there was anything leading up to this. One can rule out the Lochar Moss broken-back scrolls, because, by comparison, they are conventional if unusual for the times. But they are in openwork, and this point is important. Other openwork broken-back scrolls occur on the sword scabbard from Mortonhall, Pentlands, Edinburgh[3] (Fig. 32:5). This scabbard is included in Piggott's Group IV,[4] which is Brigantian, and it appears to be the only one with this type of decoration found north of the Wall. The rest of the scabbards in Group IV are not noted for beauty of decoration, since it tends to be simple; and without exception they come from south of the Wall. This distribution is important, since it is clear that either this Mortonhall scabbard was brought north by someone escaping from some consequence in Brigantia, or else it had been acquired under doubtful circumstances. It was brought to an area which was under the influence of the styles recently discussed; so that, were it made there, it could be expected that decoration, if any, would have been under the influence of one style or the other. Close examination shows that the broken-back scrolls are neatly enclosed by a pair of double trumpets. Some might jump to the conclusion that the presence of this decorative feature would immediately tie the object to the Elmswell-Broighter style; but in cases where these elongated trumpets have been noted, their inspiration was not local but Irish, Ireland being the land of elongated trumpets.

So, what are we back to? We are back to looking once again at origins; at the origins of all the long trumpets noted on items found in Britain, and with these we must include broken-back scrolls. We look to Lough Crew for the origins of the latter, merely because they are there in abundance, whilst elongated trumpets are everywhere on Irish bronze-work. But not one jot of this information will help us to decide whence came the broken-back scrolls on the openwork disc in the Ashmolean Museum; and the only answer possible would appear to be that these were an original idea, an effort on somebody's part to portray a naturalistic subject, to whit the branches of a tree, under the shade from which the piece may have been made. Genius will out, and for once archaeological detection is thwarted.

The broken-back idea was extended to triskeles; and this time it was a British idea, or, to narrow it down still further, possibly a Welsh idea, with some help from Galloway. The description applies to triskeles said to have drooping ends; but the ends are drooping because the back is broken. These triskeles are represented only in repoussé style, and this style alone gives them a remote relationship with the Galloway style. The obvious similarities between the two styles will be noted, and these extend even to the manner in which the work has been carried out.

In his description of the pattern[5] shown in Fig. 22:3, Megaw[6] speaks of comma tails, which he says occur on all three arms, drooping sharply from an angled break. This description is not wholly satisfactory, since, as with the Ashmolean disc, the drooping ends are really part of a devolved pattern based on the cable motif. Should this statement be thought inconclusive, then note should be taken of the decoration which occurs on a scrap of metal from a large bronze mount which is preserved at Alnwick Castle,[7] and here shown in Fig. 22:1. In this there is something more than a similarity to the much duplicated but simple pattern on the bronze collar from Llandyssul (Fig. 16:10), the essential difference being that the place of the boss on the collar is occupied by a spiral on the other. For this reason, it is possible that the Alnwick Castle piece might belong to the pre-boss period. The Alnwick Castle design is very cleverly made up of three vestigial cable motifs, two being paired, as on the Nijmegen mirror, the Polden Hill trace-hook and the Llyn Cerrig Bach plaque. A third cable motif is continuous with the spiral; so that the spiral is not so much a spiral as it is the incurved end of a cable motif, resembling the form current in an earlier age on the Continent (Fig. 16:1 and 3), and as it endeavours to be on the Turoe stone.[8] This motif on the Alnwick Castle piece is particularly fine and undevolved, and has the assurance of its being nearer to its continental counterpart than is the case with others occurring in Britain. It served as the pattern from which others were copied. Its influence is noticeable in the openwork scabbard mount from Brough, Cumbria[9] (Fig. 26:6). Post-Invasion art tends to be very abstract, but what we have here is simple and straightforward, and therefore datable to the first half of the first century. Instead of the supposed comma tails, these drooping ends of the Tal-y-llyn triskele are really part of a pattern based on that at Alnwick Castle. All the ingredients are there, and it is an established style, albeit a very secondary style, which flourished for a short time, but left no lasting memory.

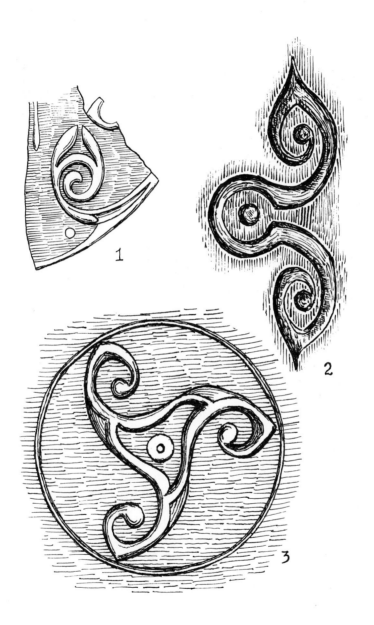

Fig. 22: 1. Decorative Detail on Bronze Mount, Preserved at Alnwick Castle. 2. Wood Carving, Lochlee Crannog, Tarbolton, Ayr. 3. Broken-back Triskele on Shield-boss, Tal-y-llyn, Merioneth. (3/4)

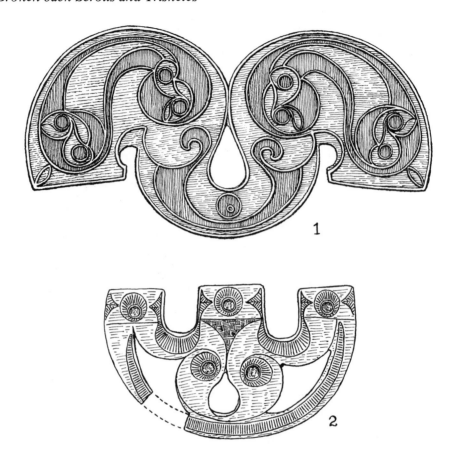

Fig. 23: Polden Hill, Somerset 1. Enamelled Mount. 2. Linear decorated Mount. (3/4)

An interesting parallel appears carved on wood. Carved wooden objects must have been quite common during these times, but are now rare because of the unstable nature of wood. Some of the wooden objects found at the Lochlee crannog, Tarbolton, Ayrshire,[10] bear simple carved patterns, one of which (Fig. 22:2) is seen to be very similar to the triskele decoration on the Tal-y-llyn shield boss. Lochlee is situated in a region covered by the Galloway style, and the occurrence here of a pattern similar in most respects to that on the Tal-y-llyn shield boss is understandable; and it makes clear, as a fact rather than as a probability, that north Wales was within an area affected by the same style; and for that reason the Tal-y-llyn shield boss could have been made farther to the north.

The Lochlee evidence is sufficient to show that the broken-back triskele form persisted at least until the second century; for the carved piece of ash was found associated with a head-stud brooch ornamented with lozenges against an enamel background, and of a type that was first made in the early second century AD. With it was a trumpet brooch, also of the early second century.

However, all influences in Wales did not emanate from Galloway. For the character of the embossed decoration on the plate from Moel Hiraddug is quite different. As Leeds[11] has pointed out, the tails that droop from an angled break at the ends of the triskele arms, again making this into a broken-back triskele, have something in common with the decoration on one of the Polden Hill mounts (Fig. 23:1). These 'tails' are in reality hanging lotus-bud motifs, and they are very much in the classical style with swollen ends; so that here, at Moel Hiraddug, there was a relationship between its decoration and that on the Polden Hill mount. It would seem that there were two influences at work in Wales, possibly dividing the country equally between north and south.

This second exercise in detection has been worth while, since, for one thing, it has demonstrated that influences were coming into Wales from the north and from the south, dividing the country into two style areas — at least at this particular time. Wales was more the recipient than the giver. Unfortunately, the detective work undertaken has not produced any explanation why scrolls and triskeles had to be broken backed in the first place. Even in Ireland such evidence is lacking. All we know is that, for a short while, everyone who devised or practiced the idea seems to have been obsessed with it.

NOTES

1. Fox, 'An Openwork Bronze Disc in the Ashmolean Museum', p. 1, pl. 1.

2. J.W. Brailsford, 'The Polden Hill Hoard, Somerset', p. 222, pl. XXIII.

3. J.M. Kemble, *Horae Ferales* (1863), pl. XVIII:5; Anderson, *Scotland in Pagan Times*, p. 120, Fig. 97.

4. S. Piggott, 'Swords and Scabbards of the British Early Iron Age', *Proc. Preh. Soc.*, XVI (1950), Fig. 9:3, A–B.

5. H.N. Savory, *Bull. Board of Celtic Studies*, 20 (1964), p. 452, Fig. 2 and pl. I:2, and II:1; *Antiquity*, XXXVIII (1964), pl. III; I.M. Stead, *Proc. Preh. Soc.*,

XXXIV (1968), p. 176, Fig. 18; Megaw, *Art of the European Iron Age*, p. 158, no. 267.

6. Megaw[5], p. 158.

7. J. Collingwood Bruce, *Catalogue of Antiquities at Alnwick Castle* (1880), p. 90, Fig. 9.

8. Duignan, 'The Turoe Stone' in Duval and Hawkes, p. 207.

9. *BM Guide to Early Iron Age Antiquities*, p. 108, Fig. 118.

10. R. Munro, *Ancient Scottish Lake Dwellings* (1882).

11. Leeds, *Celtic Ornament*, p. 56, Fig. 22.

10. Trappings

It is to be wondered just how far conceit can encourage showmanship. One is minded of the magnificently turned out mounts of the king's champion, which carried so much weight in ornamental trappings that it is a miracle it had any speed left at all. In pre-Occupation times, and perhaps for a while in early-Occupation times, trappings are known to have consisted of heavy ornamental items, such as pony-caps, of decorated bronze; bronze bridle-bits, decorated and often enamelled; harness mounts, similarly decorated and enamelled; enamelled terrets for the reins; and, for a chariot, linchpins and horn caps, as well as metal bands sometimes decorated with scrolls. The turn-out on parade must have been impressive; and this leaning towards showmanship had not been played down, even as late as the days of the Black Prince, whose appurtenances may be seen in Canterbury Cathedral.

Actually, all this ostentation, so typical of an heroic age, in no way lowered the practicability of the trappings themselves. Each and all were functional, and even the harness mounts may have had a practical purpose. A point striking to the eye is in the lavish use that has been made of brightly coloured enamels. The lavish use of enamel may have been the result of Belgic influence.[1]

Naturally, the most striking item is the Torrs pony-cap, which has been exhaustively studied by Atkinson and Piggott.[2] This is of beaten bronze, fitting snugly to the crown of the head, and it is decorated in repoussé style with motifs which have been discussed already, in foregoing chapters. Repetition would be out of place here and unnecessary; so it is proposed that the present discussion should begin with the harness mounts, which, after the pony-cap, must be the most spectacular of trappings. The most common shape is the quadrilobe form. Quadrilobe mounts have been found from Somerset to Suffolk, and one occurred at Traprain Law. Except for the specimen from Traprain Law, which is cast and undecorated (Fig. 26:2), all the rest are enamelled. The reds and blues of the enamels tend to attract the eye away from the linear decoration, which is of considerably more importance, and therefore demands close scrutiny.

Mounts of different shapes also occur, like the one found in the hoard of metalwork at Polden Hill, Somerset.[3] The shape of this mount (Fig. 23:1) may have been determined to some extent by the layout of the decoration which it bears. Although this decoration has been most carefully contrived and arranged, the quality of the draughtsmanship would have been better if judicious use had been made of a compass. Instead, this is freehand work. Nevertheless, a great deal of labour went into the making of this mount; practically the whole surface had to be undercut for the enamelling process.

Two motifs dictated the ultimate form of the pattern present. These are the lotus-bud motif and the cable motif. The lotus-bud motif appears in two styles here; the one

92

continental, as seen at Brunn-am-Steinfeld, and the other that more classical form first noted at St Mawgan-in-Pyder[4] (Fig. 10:2). The representation at Polden Hill is formal, and very close to the classical tradition. The cable motifs are very like those on the Llyn Cerrig plaque and the Llandyssul collar. At this period of time there was still some unity in styles, and for this reason one is tempted to regard the Polden Hill layout as a western style, represented from south Wales to Cornwall. However, one would be on soft ground and liable to sink at any moment if one persisted with this idea, because the shape of the Westhall mount lobes has been determined by the same cable motif layout. Part of the linear decoration on the Polden Hill mount is made up of two cable motifs with their thin ends entwined round a single boss. This does not compare with St Mawgan, or indeed with any of the Welsh styles, in which motifs occur singly, with their thin ends twisted round hemispherical bosses. But the makers of the Llyn Cerrig plaque and the Llandyssul collar felt that balance was lacking in the design, so they added a low profile embossed arc that gives to the pattern a sort of boss-and-petal look. This same idea was noted on the Turoe Stone,[5] showing that the Welsh metalworkers were following a precedent. Now, this balancing-out trick is noticeable also amongst the Polden Hill decoration, giving it too a boss-and-petal look; but neither of the two patterns on the St Mawgan and Westhall mounts has anything that is comparable, each being of a form that is common to both. Therefore, the style of this decoration on the Polden Hill mount is possibly Welsh, or the Welsh style was suggested by Polden Hill; whereas Westhall shares a common style with St Mawgan.

A second mount from Polden Hill is quadrilobe, like the majority, and it exists in strong contrast to the first. The decorative style (Fig. 24:3) is more abstract. Also it must be quite a deal later. It has an edging of small triangles, which are enamel filled, and all pretence at a linear pattern has been abandoned. The rather crude S-shaped designs, against an enamel background, are reminiscent of others on the first mount, and again on the Desborough mirror, and they are based on the non-boss-and-petal pattern, common in Wales. The idea is better expressed on the first mount. Devolved art forms are never easy to analyse. Equally, the design could be a devolved form of that seen on the bronze plaque from Knobb's Crook, Woodlands, Dorset[6] (Fig. 26:4). The arc-shaped side panels enclose other devolved motifs, which could be either palmettes or lotus-buds.

The third mount found in the Polden Hill hoard of metal objects (Fig. 23:2) is decorated entirely in the linear style, which here is very reminiscent of mirror decoration. But the overall design, with its sweeping curves, bears a strong resemblance to the general layout of the decoration on the first mount, which Cyril Fox[7] thought of as a horse brooch. It is becoming increasingly clear that the mirror style, which this third mount imitates, was first suggested by the form of the decoration on the first mount; and the continuing connection is well expressed by the basketry filling on the third mount. The mirror style, which is a linear style, was a quick and easy method of decorating a large area of metal, such as that presented by the reverse sides of the mirrors. The sweeping curves of the first mount served as a basis for those on the mirrors — largely imitations of them, with variations. In general, the style of the decoration on these Polden Hill mounts would date the lot to the middle of the first century AD, with the first mount being anything up to a decade earlier. The source of the mirror style can be looked for on this first mount. An important site indeed is Polden Hill.

Westhall, which is three miles north-east of Halesworth, Suffolk, is noteworthy as being the location of a hoard that was brought to light more than a century ago. The hoard is also noteworthy for the quantity and the quality of the objects unearthed. Eight enamelled terrets, linchpins, a matching pair of quadrilobe harness mounts enamelled in green, white, blue and red, were found inside a thin bronze vessel, on which there was an embossed cruciform pattern with palm branches between the arms; and, according to an early account,[8] associated with it was a coin of Faustina the Elder (AD 138–41). This is a dubious association, since, as Rainbird Clarke[9] has pointed out, these objects are most likely to have been concealed *c*. AD 60, when the Boudiccan Rebellion[10] was put down by the Romans amidst terrible slaughter. All objects are of early types and they are unlikely to be later than the latter end of the first half of the first century AD. A disc-brooch, also found with the hoard, has Belgic affinities. In view of the quality of the decoration, this hoard clearly indicates the spread of Belgic craftsmanship to East Anglia, some time after the adoption of polychromy in the phase of Claudian Romanisation.[11]

Fig. 24: Enamelled and Openwork Harness-mounts 1. and 4. Santon, Norfolk. 2. Westhall, Suffolk. 3. Polden Hill, Somerset. 5. Stanwick, Yorkshire. (1/2) except 1 (3/4)

The multicoloured impression left on the mind by the pair of matching enamelled mounts (one of which is shown in Fig. 24:2) must not blind us to the carelessness of the workmanship. This again is the result of bad draughtsmanship. The left-hand pattern is repeated on the right-hand side, but minor variations are obvious. Irregular chiselling was a contributory cause also. The enamel background throws into relief a trefoil pattern with centre boss, and enclosed within a circle in the middle of the mount. To right and left of it are degenerate broken-back scrolls, the ends of which are expanded into abstract boss-and-petal patterns. There is a deal of muddled thinking relative to these broken-back scrolls, indicating a condition of remoteness, which is understandable, since the broken-back scroll is essentially a western style. Due to a lack of understanding, a pattern emerged which was repeated at Stanwick (Fig. 24:5) and at Eckford, Roxburghshire[12] (Fig. 26:7). But this new conception did not originate with the Westhall mount: it is seen to perfection and better drawn, better executed, on the mount from Santon (Fig. 24:1). The Westhall mount decoration is generally rather devolved, making it probably one of the last products to come out of Icenia before the Boudiccan Rebellion of AD 60. Yet the linear decoration, such as it is, does not really support this view. Cable motifs decorate the lobes, though sometimes they are carelessly represented. The lobes themselves have been shaped as they are in order to accommodate them. The style is very similar to the linear pattern on the St Mawgan shield mount, even to the inclusion of non-bosses. Also, there are elongated cable motifs of the form appearing on the Llyn Cerrig plaque, where, as here, they enclose the pattern. Thus an anomaly is present, in that an early style is coupled with a devolved one; and the anomaly can only be explained away by somebody having a long memory.

Similarities occur amongst these mounts: they are strong, and they provide evidence of a common approach to articles of this nature throughout East Anglia, with some exerted influence to do the same in the south-west, suggested by the matching forms of the second of the Santon mounts with the second Polden Hill mount. This causes us to examine the quadrilobe form more closely, and we find that it is nothing more than two Polden Hill mounts of the first form (Fig. 23:1) set back to back. There is no mystery about this: it makes for a perfectly balanced design. The determination of lobe shape had to be in accord with the manipulated cable motif, as seen in Fig. 24:1, which is a mount from Santon. Here the workmanship is of the finest quality, strikingly imaginative, and undoubtedly of Belgic origin. By comparison, the second Santon mount (Fig. 24:4) is plain and uninteresting. Such is the decoration on this second mount that it must proclaim a partial collapse in pattern-making procedures, perhaps occasioned by the disturbed conditions of the times. Inherited motifs do not get the attention they deserved.

The Santon hoard[13] of trappings was packed into a bronze cauldron.[14] Its contents may be classified into two groups: objects of native workmanship, and Roman imports from the Continent. Smith's conclusion[15] that the hoard illustrates the Romanisation of Britain at the expense of native traditions is interesting, but it is hardly warranted by the evidence. A Claudian–Neronian date is suggested for the hoard because of the presence of Roman thistle brooches and the jug, all of which are most likely to have been imported prior to the Claudian Invasion. Like the Westhall hoard, this one was also probably buried *c*. AD 60. In addition to the mounts mentioned, the hoard contained, amongst the native objects, a two-link bridle-bit, nine linchpin heads, five bronze axle ends or ferrules, and six bronze nave ends.

Megaw[16] maintains that the first of the Santon mounts (Fig. 24:1) is not so elaborately decorated as is a badly preserved quadrilobe specimen from the London area, which is decorated in red and yellow enamel. He may have arrived at this conclusion because most illustrations of the Santon mount minimise its most important feature, which is its linear decoration. Coloured illustrations tend to seduce the eye away from this linear decoration to the coloured enamels; but it is the linear decoration which alone provides that information necessary to a complete understanding of the rightful place of this object in British art. And when this linear decoration is studied, it will be seen that the whole basis of it is a proliferation of cable motifs. There is a singleness of thought here, in that the artist's chief idea has been to evolve as complicated a design as possible, using only one motif. In this respect, he has succeeded magnificently. There is nothing else quite like this in the whole of British art. The artist was assisted throughout by a pair of compasses, of which he made full use; and because he did so, there is perfect symmetry. No area, not even a corner of the surface of the metal remains untouched by his scriber. Note how well the cable motifs have been fitted into the spaces available on the lobes, which were shaped like this in order to accommodate them. Some motifs have been stretched out: but what should have been their thin ends bent into a coil, are here expanded to contain linear loops with swollen ends in the Irish manner. As a result of this treatment, there is a sort of ornithomorphic look to some of this decoration, which may not be intentional; but the whole idea is repeated at the ends of the broken-back scrolls which occur on the side panels. The ends are shaped neither like the motifs on the Llandyssul collar, nor like those on the St Mawgan shield mount, but are examples of native craftsmanship under Belgic influence *c*. AD 40–60, as Clarke has pointed out.[17] Thus, we have here an entirely new version of the broken-back scroll, which differs completely from the form which was introduced into Britain from Ireland. Here, on the east coast, connections with Yorkshire must have been close, since the very same broken-back scrolls occur, but in openwork form, on the Stanwick mounts (Fig. 24:5), and again on a cheek-piece from Eckford, Roxburghshire (Fig. 26:7), which is a town situated in the territory of the Votadini.

On both the Santon and Westhall mounts some cable motifs have been stretched out to look like elongated trumpets. The differences between these and the trumpets are to be found in their blunt swollen points, whereas elongated trumpets have moulded seed-shaped ends. There is not likely to be any great difference in date between these two mounts. Bosses are still unknown here, so their places are occupied by a round of enamel, though still suggesting a non-boss style. At both Santon and Polden Hill quadrilobe mounts were made, or was there some system of exchange? The simple decorative styles owe nothing to the cable motif. But there seems to be an age difference. Linear decoration was forgotten, or it had had its day: instead, attention is now directed to the enamel background, which throws into relief the patterns we are meant to look at. Undoubtedly, both are post-Invasion, and were made some time between AD 43–60. The more imposing examples are certainly pre-Invasion; but, because they were considered valuable, they got carried over into a later period, and all types ended up by being buried together.

Mention must be made here of the quadrilobe openwork harness mount from Norton,[18] which bears a certain resemblance to the Westhall specimen. It is enamelled in red and yellow dots, with stippled spaces, indicating some Roman influence. The Traprain Law[19] harness mount (Fig. 26:2) is a badly cast quadrilobe affair carried out in boss-and-petal style, but there is no enamel. Traprain Law was the home of the boss-and-petal pattern. Another mount rather like it comes from Middlebie, Dumfriesshire.[20] The style reached a state of rapid devolution in the north. Clearly, the quadrilobe form had moved into alien territory, where, even at Traprain Law, the oddly shaped side members of this mount, springing as they do from a central boss, appear to have defeated the metalworker as well as us; for the layout calls for no ready explanation, though it is quite pleasing to the eye.

The Stanwick style[21] is something of a rarity. Here an individual style in openwork decoration was developed with the help of a strong stimulus from the south-east. This came in the form of broken-back scrolls, based on the cable motif, so aptly expressed on the Santon harness mount: a linear pattern that encouraged the metalworkers to chip away on all sides for an enamel background. The metal that was left proud in this sea of enamel was then translated into openwork, as at Stanwick, and this was bound to an outer ring with metal clips resembling the 'lips' of lipped terrets.

Lipped terrets shock by their very ugliness. The Celtic mind, so used to observing well balanced designs and sweeping curves, sometimes succumbed to curious deviations. The mood was reactionary. The reaction at Stanwick was mild compared with elsewhere: the lipped horrors that paraded as terrets are almost dinosaurian. Leeds saw the application of lips as a conceit of western artificers, where the most monstrous efforts are found. But the idea may have originated in the eastern part of the country, where these lips are more restrained. There are four, and perhaps more localities in East Anglia, where these terrets have been found, and we know they occurred at Stanwick, where lips appear to have been adopted as a local trademark, covering most of the items produced there. A far northern occurrence was at Newstead.[22] Two were found in Sussex; one at Alfriston,[23] and one on Wilmington Down.[24] The rest came from Polden Hill and Knowle, Somerset,[25] and from thence up the Severn to Wroxeter, with a stray occurring at Ditchley, Oxon.[26] Amongst the East Anglian terrets there are two lipped examples which came from the Stanton hoard, one of which is illustrated in Fig. 25:4. Inclusion in this hoard suggests a *terminus post quem* date for this terret of AD 60. Taken together, there is nothing very remarkable about the forms of these terrets: like the Stanton specimen, often the lips are very restrained, but against this there are the excesses of Polden Hill. One terret is shown in Fig. 25:5, and it will be seen to be exuberant in its aggressiveness, with grossly enlarged lips (tartish is the word suggested) each having petal-shaped inserts of enamel, whilst triangles of enamel decorate the hoop. This large-lipped terret can hardly be later than the one from Stanton, yet the differences in approach are remarkable.

Terrets are rings through which the reins were passed. Generally, this ring is not circular, but instead it is slightly elliptical, whilst at the base, where it is attached to the harness, the ring is reduced or enlarged to a thin rectangular plate. In lipped terrets the hoop is of even cross-sectional area. On the other hand, there are plain terrets in which the cross-sectional area shows a steady increase from top to base, the expansion in girth here being anything up to twice the cross-sectional area of the top. One of these plain terrets is shown in Fig. 25:1. It comes from Polden Hill, an area subjected to a number of cross-currents. Fox[27] has demonstrated how these plain terrets received further additions, such as knobs and enamelled plates, the latter being referred to as the 'crescent' type, from the shape of its decorated plate. For these, the Belgae are said to have been responsible. Although normally passed over without much consideration being given to their form, nevertheless there is an intelligent basis for the shape of the plain terret: they are made up of two conjoined trumpets, with the trumpet mouths engaged in a bar-swallowing act — the bar being for attachment to the harness. This plain form had a long life, from perhaps just BC to as late as the beginning of the second century AD. One of these plain

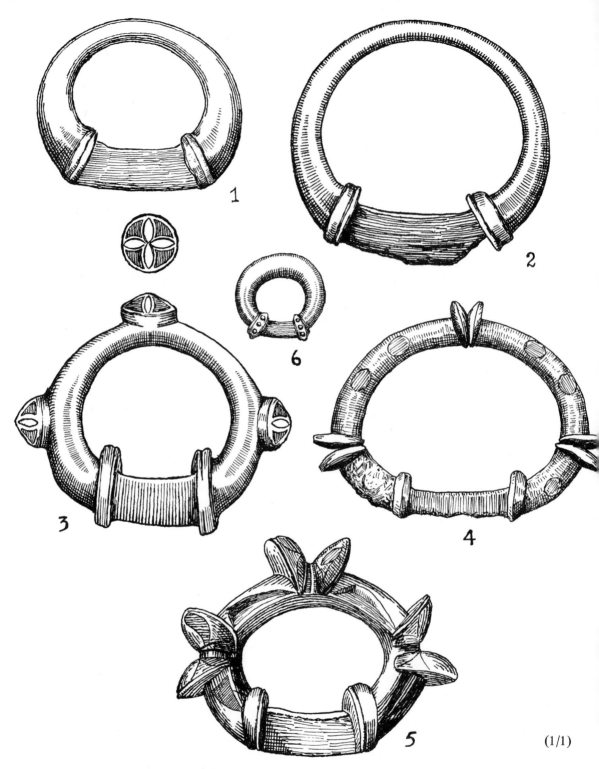

Fig. 25: 1. Plain Terret, Polden Hill, Somerset. 2. Plain Terret, Llyn Cerrig Bach, Anglesey. 3. Decorated Knobbed Terret, Seven Sisters, Neath, Glamorgan. 4. Enamelled Lipped Terret, Stanton, Suffolk. 5. Enamelled Lipped Terret, Polden Hill, Somerset. 6. Miniature Terret, Trevelgue, Newquay, Cornwall.

(1/1)

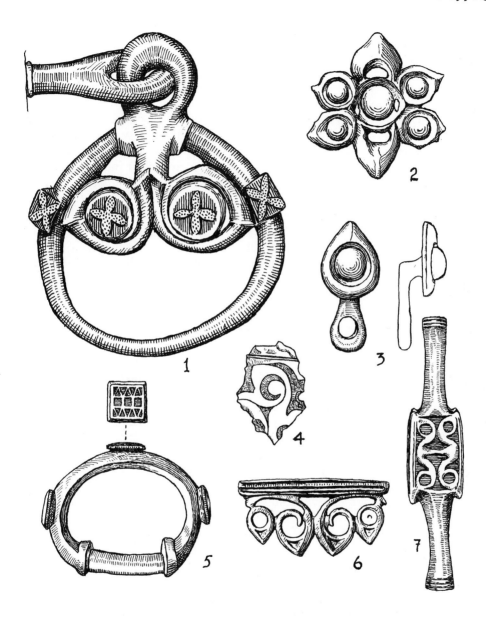

Fig. 26: 1. Enamelled Bronze Bridle-bit, Holderness, E. Yorks. 2. Harness Mount in Boss Style, Traprain Law. 3. Boss-and-Petal Dress-fastener, Colchester, Essex. 4. Detail from Decoration on Bronze Plaque, Knobb's Crook, Woodlands, Dorset. 5. Enamelled Platform Terret, Birrens, Dumfries. 6. Bronze Openwork Scabbard Mount, Brough, Westmorland. 7. Enamelled Bronze Cheek-piece, Eckford, Roxburghshire. (3/4) except 4 (3/8)

terrets was found at Llyn Cerrig Bach.[28] They were in the Middlebie hoard,[29] and one was found at Torwoodlee, Selkirkshire.[30] There is an extraordinary congregation of them between Bridgewater Bay and Poole Bay in south Britain. Miniature plain terrets (their purpose can only by guessed at) have been found on several British sites: that shown in Fig. 25:6 is from the promontory fort at Trevelgue, near Newquay, Cornwall.[31] This site also yielded a linchpin of Yorkshire type.

The plain form remained virtually unaltered; and even when it became the recipient of knobs and crescents its form still remained the same. The majority of the knobbed terrets are clustered together in what John Gillam has termed the Inter-Isthmus Province,[32] with a slight overspill into Brigantia. Rarely does one get such a clear indication of where the workshops were. Some others have been found in south Wales and the Severn Valley — another pointer to regional production. Most of these terrets have come from native sites, but several also have been found in forts and milecastles along Hadrian's Wall. But, in all these cases the knobs are plain. The Silures in south Wales owned a pair of terrets at Seven Sisters, Glamorgan,[33] which were ornamented with enamelled designs in two colours, the design being a quatrefoil one with subtriangular fillings. One of these Seven Sisters terrets is shown in Fig. 25:3. These terrets are slightly out of place in a hoard which contained objects like a bridle-bit with cellular decoration, and other items on which the decoration is characteristically Roman. On balance, these are post-Invasion items. Perhaps most belong to that anomalous period which preceded the building of the Wall.

By contrast, the distribution of crescent terrets is predominantly East Anglian, with an overspill into the Thames Valley and Kent. The most spectacular of these are shown in Fig. 27. Strays have been reported from Auchendolly, Kirkcudbright;[34] Pentyrch, Glamorgan;[35] and from Wiltshire. The nature of the East Anglian discoveries leaves little room for doubt but that most, if not all, were buried *c*. AD 60, after the failure of the Boudiccan Rebellion.

Of the five terrets illustrated in Fig. 27, three come from Westhall.[36] All have been illustrated before. The basis of the terret form is the twin trumpet, with added crescent. This crescent is used as a vehicle for decoration, in this case in the form of palmettes and scrolls. The fact that the largest Westhall terret bears a row of small triangles at the base of the crescent is a reminder of similar occurrences on the Trelan Bahow mirror and on the quadrilobe mount from Polden Hill. The period of manufacture of these terrets can safely be put somewhere between AD 43–60, and almost certainly they are the products of Icenian workshops. From the beginning, the Iceni had supported Rome, and their wealthy king, Prasutagus, considered himself fortunate in backing the winning side. But, as a result of the Fosse Way Limes[37] commotion in 47–8, the Iceni became disillusioned at what they regarded as 'encirclement',[38] and there followed the disarmament of themselves and of their kinsmen. The crushing of the revolt by Roman auxiliaries under Ostorius Scapula may have brought about the hiding of hoards of objects which had been proscribed.

On the largest Westhall terret the decoration is simple enough. At the centre of the crescent there is a simplified and highly stylised palmette. To the right and left of this the cable motif has determined the form of the remaining part of the pattern. The same overall design is repeated on the smaller terret, Fig. 27:2. On the third and smallest terret the decoration is highly stylised, with little meaning. The motifs represented are very much devolved, giving an air of degeneracy to the pattern as a whole. In all cases

Fig. 27: Crescent Terrets, Enamelled 1.–3. Westhall, Suffolk. 4. Whaplode Drove, Lincs. 5. Lesser Garth, Pentyrch, Glamorgan. (3/4)

the terrets are enamelled, the enamel forming a background to the patterns. Altogether, there were eight terrets in the Westhall hoard.

But, if the decoration is considered degenerate here, so it was elsewhere as well, on other terrets of the same form, such as those from Auchendolly and Bapchild. The formless pattern on the Auchendolly terret[39] incorporates but the shadow of a simplified palmette. On the other hand, the decoration on the Bapchild terret[40] consists simply of running simplified palmettes. Though of degenerate character, there is no doubt but that this was a style, a real decorative style which was current in the mid first century AD, and it was by no means localised. The process of degeneration was carried further in the case of the crescent terret (Fig. 27:4) from Whaplode Drove, south Lincolnshire.[41] For decoration, there are two sets of what might be termed arc-and-three-circles; but the meaning of this symbolism is lost. The coin evidence at Whaplode Drove suggests a late-first- or early-second-century date for this terret. It has been subjected to long usage, and for that reason the date of its manufacture could be put back a little. In Aileen Fox's study[42] of the Richborough terret, it has what she describes as 'a continuous scroll pattern of late Celtic type'. The centre motif is clearly palmette based, whilst the remainder of the pattern could have been derived from Westhall.

An unusual terret comes from Lesser Garth, Pentyrch, Glamorgan (Fig. 27:5). Maybe it was wrong to describe this terret as a stray, because it is very much in the nature of a hybrid, and it seems very much at home in south Wales. It might have resulted from a local attempt to combine two forms, by drawing on ideas from Westhall and Seven Sisters. The three bosses indicate knobbed terret influence, whilst the rather reduced size of the crescent shows a lessening of eastern influence. The curvilinear decoration of red enamel outlined with dots clearly shows Belgic or Icenian influence, though there is every reason for agreeing with Savory[43] when he states that this is the work of south Wales metalworkers, the link with the Belgic south-east being the associated linchpin and the cauldron hanger. The decorated bosses have been cast in two pieces, an interesting variant to normal practice. The top halves are decorated with red enamel dots and small spherical triangles, set in a regular pattern. A date for this terret is suggested by the accompanying iron work, which is thought to antedate the Roman Invasion of south Wales.

Belgic influence has been clear in the enamelled pieces discussed above. The champlevé enamel process, common to all, first made its appearance in southern Britain on trappings in known Belgic areas. The near continental nature of some of the patterns used is particularly evident in the form taken by the palmettes on the scabbard top from Verulamium, St Albans, Hertfordshire (Fig. 12:4). The style is similar to that expressed at Westhall. The spread of Belgic techniques to other parts of Britain is a useful tribute to Belgic influence. There is nothing in the process of enamelling that could not be picked up quickly by any competent metalworker; yet, with perhaps the sole exception of Polden Hill, the best pieces are still to be found in eastern Britain.

The history of bridle-bits began long before the first century AD, and, for that reason, early development is not our concern here. Ward Perkins[44] has drawn attention to the dual nature of bridle-bit construction in early Iron Age Britain: but, in the present instance, immediate interest is in bridle-bits made up of three elements — two rings with excrescent bars and a connecting link — since this form comes well within our period. It is also the form which received the maximum of decoration. Looked at from another angle, there are seen to be three links, in which the outer links are drawn within the circle of rings intended for the reins: these rings are attached to the outer links about their middles.

When the outer links penetrate within the circle represented by the rings, they are decorated, sometimes quite elaborately. Also, there is the practice of decorating one with more elaborate decoration than is the case with the other: the reason for some bits having dissimilarly ornamented rings being that these bridle-bits formed part of a double harness for a pair of horses, in which the more elaborately decorated rings were situated on the outside. Examples: a bridle-bit intended for a one-horse harness is that from the Lochlee crannog, Tarbolton, Ayrshire;[45] whilst one intended for a double harness suitable for a pair of horses is that from Middlebie, Dumfriesshire[46] (the outer ring is shown in Fig. 56:1).

Certain styles of decoration were applied to bridle-bits, and they are not always the same. Some are fairly localised. On the Middlebie specimen there is both a single- and a double-boss style of decoration. This style occurs on dress-fasteners, like those from Glastonbury;[47] but the same style of fastener is also noted at Lochspouts crannog, Maybole (Fig. 45:4 and 12) where it is dated as late as the second century. But at Lochspouts it is also associated with a double trumpet pattern, which is seen to occur on the cheek ring of a bridle-bit from Chesterholm, Northumberland.[48] The boss-and-petal pattern, so common at Traprain Law, especially on dress-fasteners, is represented on two fragments of a bridle-bit from Newstead, Roxburghshire.[49] There will always be others, of course, but these instances are sufficient to demonstrate that no particular form of decoration was reserved for bridle-bits, and that they were decorated with whatever motifs and patterns happened to be current at the time of manufacture.

From the style of this decoration, these bridle-bits can be dated well into the Occupation period, but probably not later than around the second half of the first century. But there are others which, by reference to their decoration, may be looked upon as being earlier, like those from Holderness, east Yorkshire[50] (Fig. 26:1) and the Thames at London (Fig. 15:4). The lozenge-decorated Greek cross of the latter suggests a post-Invasion date for its manufacture, but the rounded trumpets are reasonably early, so that a mid-first-century date would seem to be appropriate for this piece. The Holderness bridle-bit is probably of much the same period, but definitely before AD 74. Its trumpets, wrapped as they are around quatrefoil ornamented non-bosses, make up a pattern that is ancestral to the true boss-and-petal style. The entwined continental trumpets of the Thames bridle-bit are perhaps a little earlier than are those on the Holderness bit, and they are in keeping with the spherical triangle which decorates the end of the link. The Holderness trumpets are more evolutive, and are undoubtedly Brigantian. There is less enamel in the north, and this is noticeable at Holderness. Proceeding yet further north, it will have been noted that the Scottish and Northumberland specimens quoted above are entirely without enamel.

Plainness results in a situation of stark realism; and this factor is noticeable particularly in the case of the pair of mounts (usage unknown) found in association with a massive armlet and a Roman patera at Stanhope, Stobo, Peebles.[51] Here the stark realism of a repetitive boss-and-petal pattern (Fig. 56:8), which somehow must express a bankruptcy of artistic thought, and an absence of suitable patterns, may have resulted from alien pressures. These objects were not very old when they were buried with the massive armlet and the Roman patera. For this reason they are perhaps later than any other object so far described. As to the pattern itself, there is a reflection of this on an embossed disc from Annalore, Co. Monaghan (Fig. 33:1). The boss-and-petal is perhaps the last of the truly Celtic-inspired patterns, and its history may have been prolonged for that very reason.

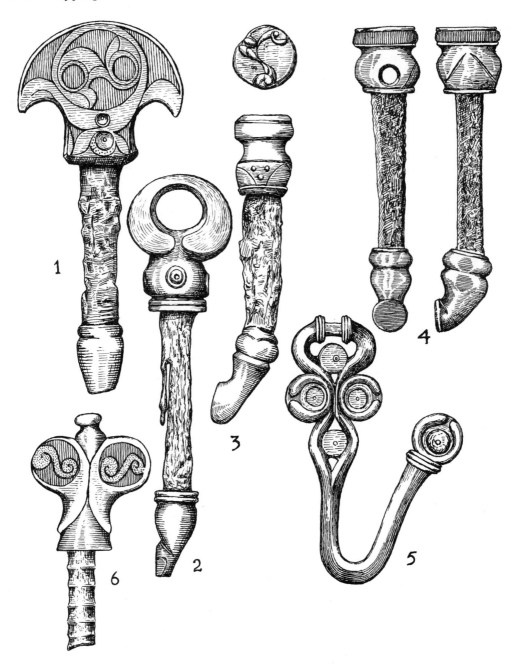

Fig. 28: Linchpins, of Iron with Bronze Ends 1. Enamelled, King's Langley, Herts. 2. Middleton, E.R. Yorkshire. 3. Wigginton Common, Tring, Herts. 4. Llyn Cerrig Bach, Anglesey. 6. Enamelled, Camulodunum, Essex. 5. Trace-hook, Polden Hill, Somerset. (3/4)

Linchpins are used to retain the wheel on the wheel-shaft. In our period they were of iron with bronze heads and bases; and the bases were slightly tapered to prevent their working free from the shaft-hole. The bases leave little room for decoration, so there is · none. Heads, on the other hand, are occasionally decorated. In his study of linchpins, Ward Perkins[52] made out a case for there being two major groups — the south-east type, and the Yorkshire type. Both have a respectable continental ancestry, but those histories are not for us, since they antedate our period. Most of the British decorated examples come within the first century, and therefore they are of interest here. Some patterns can be quite elaborate, like that on the Yorkshire-type specimen (Fig. 28:3) from Wigginton Common, near Tring, Hertfordshire.[53] One would imagine that decoration on a linchpin would have been regarded as wasted effort.

The Yorkshire type is characterised mainly by the shape of its base. As a group, the Yorkshire type is very consistent, and it had a long life. The base or foot, is lipped, and the general pattern and head shape show hardly any variation.Ward Perkins claims a Marnian ancestry for them, with the likelihood that they were brought to Yorkshire by Marnian invaders, of whose chariot burials they were a characteristic feature. Nevertheless, it must be pointed out that what is called the Yorkshire type of linchpin also occurs in Essex, Hertfordshire, Norfolk, Kent, Somerset, Cornwall and Hereford. Four specimens were found at Colchester in area A, not stratified, but abandoned *c*. AD 65. Others were found in the mid-first-century-AD hoard at Stanwick; and again in hoards of similar date at Santon, Norfolk,[54] and at Polden Hill, Somerset.[55] In the Santon hoard there was also a linchpin of the south-east type,[56] and another came from Boudiccan levels at Colchester.[57] The south-east type included a number of plain iron pins, notably from Brading, in the Isle of Wight; Chedworth, Gloucestershire; Cirencester (four are in the Corinium Museum); Maiden Castle, Dorset; and finally from London. The south-east type has also been found in Yorkshire.

Probably, the most westerly occurrence of the Yorkshire type is at Llyn Cerrig Bach (Fig. 28:4). The hoard included tires, nave hoops, a bronze horn-cap, bridle-bits, terrets — enough to make possible the reconstruction of a chariot. The people who had hidden this hoard may have been refugees from the Belgic areas, fleeing before the Roman advance.[58] The linchpin from Wigginton Common is probably the earliest of those illustrated in Fig. 28; for the style of head decoration is not unlike that on the gold torcs and the Ulceby terret, as well as the Llyn Cerrig plaque. The dominant pattern is the cable motif. The same motif dominates the design on the King's Langley, Hertfordshire, linchpin (Fig. 28:1); but now the decoration is linear, and there is an enamel background. Part of the decoration seems to be anticipating the boss-and-petal pattern. Situated at the base of the head is a double cable motif associated with a non-boss. This is a most unusual form of head, of a shape intended to provide a good finger grip. Belgic influence is strong here in this pattern, and the style is pre-Invasion.

Of Neronian date is the curious looking linchpin which was found in the Boudiccan destruction debris in Region 3 over road II at Colchester.[59] Here the iron pin is ribbed. The lateral lobes are faced with roundels of champlevé enamel, with blue S-scrolls on a red background. The S-scrolls should be compared with those on the Polden Hill mount (Fig. 24:3). Another linchpin with an unusual head shape is the specimen found between Middleton and Enthorpe, in the east Riding of Yorkshire (Fig. 28:2). The cast head has in relief a design which imitates that in penannular form on the trace-hook from Polden Hill (Fig. 28:5). Appearing at Polden Hill in triplicate, the design is derived from the devolved double cable motif seen on the openwork disc in the Ashmolean Museum (Fig. 17:6) which, with its broken-back scrolls is of pre-Invasion date. An uncommon British design, nevertheless its beginnings go right back to the double-cable motif seen at Schwarzenbach, Germany (Fig. 16:2), which is of fifth-century-BC date.

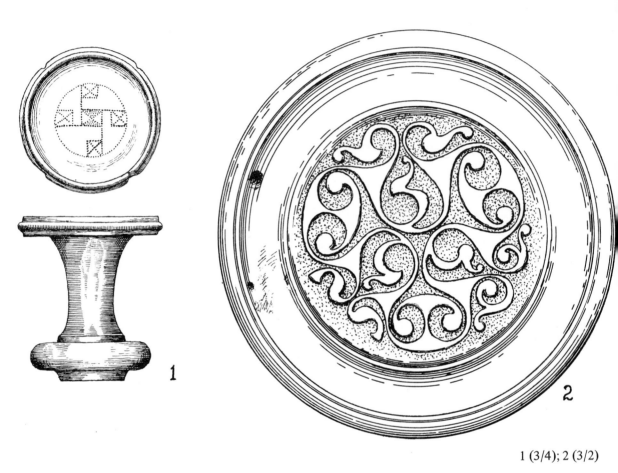

1 (3/4); 2 (3/2)

Fig. 29: Decorated Bronze Horn-caps 1. Llyn Cerrig Bach, Anglesey. 2. Brentford, Middlesex.

Finally, in this round-up of trappings, something must be said about horn-caps, which were mounted on the tops of the chariot-handles. Some have been turned on a lathe, whilst others appear to have been made out of spun metal. All are not made in one piece: the top and bottom mouldings being added to the central section. There is a southern emphasis to the distribution, and, curiously, the form is unknown in Brigantia. Three were found in the River Thames, at Brentford,[60] Putney[61] and at Goring. Two have come from Dorset, at Maiden Castle[62] and Horton;[63] whilst others have come from Bigbury, Kent;[64] Burwell Fen, Cambridgeshire;[65] High Cross, Leicestershire;[66] Ham Hill, Somerset;[67] and Llyn Cerrig Bach, Anglesey.[68] The last named is illustrated in Fig. 29:1: its swastika-type decoration can be paralleled on the Battersea shield.[69] This Llyn Cerrig horn-cap is typologically an early specimen, with a mid-eastern origin for its decoration. The only dated specimen is the typologically later horn-cap from Maiden Castle, which Wheeler dates to the period *c.* AD 25–45.[70] It came from a Belgic level. The Putney horn-cap's decoration is reminiscent of some spoon patterns. From the purely decorative point of view, the most spectacular specimen is the horn-cap found in the Thames at Brentford, and said by Jacobsthal to be 'continental down to the smallest detail'. Other exaggerated claims have been made for it, de Navarro[71] mistakenly placing it amongst the earliest examples of Celtic Art in Britain; whilst Powell[72] thought that the decoration as a whole is typically Waldalgesheim.

Clearly, some revision of dating is called for, though in recent years there has been a move to include it in much later contexts. As an example, the compilers of the Catalogue of the Edinburgh Exhibition of Early Celtic Art[73] pointed out that the ornamentation on the Brentford horn-cap was produced by paring and pinching down the surrounding metal, as was done in the case of the bronze collar from Stichill (Fig. 19). For this reason they incline to date the horn-cap to the turn of the century BC–AD. The design in question, considering the small area available for decoration, is quite an elaborate one, and it is clearly palmette based, with the palmette here represented in its most abstract form. In fact, this was the ultimate in devolved palmette forms, and it is even more devolved than are the palmettes of the Battersea shield. Further, in this form it survived the Occupation, appearing as voids in this form on the hanging-bowl escutcheons, like those from Baginton, Eastwell, Hildersham, Tummel Bridge, Castle Tioram and Craig Phadraig. But Piggott and company's date as given above is in direct conflict with the beliefs of Otto Hermann Frey, who, by drawing attention to its 'looser' scrolls, proposes a close Waldalgesheim ancestry;[74] and with regard to the Brentford horn-cap's period of manufacture, he believes 'therefore it cannot be placed before the mid-third century'. Meaning of course, BC and not AD. There is no basis for this belief, and moreover it is completely at variance with the evidence. The Brentford horn-cap must belong to the early part of the first century, and definitely pre-AD 43; and Piggott's dating is very near the mark, as one would have expected.

NOTES

1. It is known that at some time after 75 BC, Belgic invaders had dominated Hertfordshire. They were far ahead of the British in material culture, and they waxed rich mainly because of their exploitation of rich soils neglected by the Britons. These Belgae were acquainted with the Romanised world of the Mediterranean and of Gaul, and some of them at least were probably fleeing before the Roman advance overtook them. In the early years AD they extended their dominion to include Cambridgeshire, but here they halted. Also they laid claim to the territories from Southampton Water to the Bristol Channel, touching upon Gloucestershire. The Belgae were expert enamellers, and it is clear that they had a hand in the production of the objects now to be discussed.

2. Atkinson and Piggott, 'The Torrs Chamfrein', p. 232.

3. C.J. Harford, 'Account of Antiquities found in Somerset', *Arch.*, XIV (1808), p. 90, pls. XVIII–XXII; Brailsford, 'The Polden Hill Hoard, Somerset', p. 222 and illustrations.

4. Murray Threipland, 'The Excavations at St Mawgan-in-Pyder, north Cornwall', p. 80, pl. XI and Fig. 40.

5. Duignan, 'The Turoe Stone' in Duval and Hawkes, p. 206, Fig. 4.

6. Fowler, 'A Roman Barrow at Knobb's Crook, Woodlands, Dorset', p. 36, Fig. 5:7.

7. Fox, *Pattern and Purpose*, p. 89.

8. H. Harrod, 'On Horse Trappings found at Westhall', *Arch.*, XXXVI (1856), p. 454, pl. XXXVII.

9. Clarke, 'The Iron Age in Norfolk and Suffolk', p. 69.

10. The date of the Boudiccan uprising is now reliably dated to AD 60. See Dudley and Webster, *The Rebellion of Boudicca* (Routledge, 1962).

11. F. Henry, 'Emailleurs d'Occident', *Préhistoire*, II, p. 103.

12. Curle, 'Objects of Roman and Provincial Roman Origin found in Scotland', p. 365, Fig. 49.

13. R.A. Smith, 'A Hoard of Metal found at Santon, Norfolk', *Cambs. Antiq. Soc. Proc.*, XIII (1909), pp. 146–63; *Arch. Journ.*, XCVI (1939), p. 69; *Antiq. Journ.*, XX (1940), pp. 341ff.

14. *Proc. Preh. Soc.*, V (1939), pp. 175–6.

15. R.A. Smith in *V.C.H. Suffolk*, I (1911), p. 321.

16. Megaw, *Art of the European Iron Age*, p. 163.

17. Clarke[9], p. 69.

18. Kemble, *Horae Ferales* (1863), pl. XIX:2 and 4.

19. Curle and Cree, 'Excavations at Traprain Law', *Proc. Soc. Antiq. Scot.*, L (1915-16), p. 112, Fig. 28:2. A somewhat similar one came from Middlebie, Annandale – *Journ. Roman Studies*, III, pl. II – whilst more simple forms in the same style are known from Newstead – Curle, *A Roman Frontier Post and its People*, p. 302, pl. LXXV:1 and 3.

20. J. Curle, 'Roman and Native Remains in Caledonia', *Journ. Roman Studies*, III, p. 99, pl. II.

21. M. Macgregor, 'The Early Iron Age Metalwork Hoard from Stanwick, North Riding, Yorkshire', *Proc. Preh. Soc.*, XXVIII (1962), p. 17.

22. Curle[19], p. 302.

23. Kemble[18], p. 196 and pl. XX:2.

24. *Arch. Journ.*, X (1853), p. 259.

25. A Bulleid and H. St George Gray, *The Glastonbury Lake Village* (Glastonbury Antiq. Soc.), I, p. 230, Fig. 45:A.

26. C.A.R. Radford, 'The Roman Villa at Ditchley, Oxfordshore', *Oxoniensia*, I, p. 55, Fig. 10 and pl. X.

27. Fox, *A Find of the Early Iron Age from Llyn Cerrig Bach, Anglesey*, p. 28.

28. Ibid., no. 44.

29. Wilson, *Prehistoric Annals of Scotland* (1863), p. 458 and illustration.

30. J. Curle, 'Notes on two Brochs recently discovered at Bow, Midlothian, and Torwoodlee, Selkirkshire', *Proc. Soc. Antiq. Scot.*, XXVI (1891–2), p. 80, Fig. 10.

31. J.B. Ward Perkins, 'An Iron Age Linch Pin of Yorkshire Type from Cornwall', *Antiq. Journ.*, XXI (1941), p. 64, pl. X.

32. J.P. Gillam, 'Roman and Native A.D. 122–147' in Richmond (ed.), *Roman and Native in North Britain*, pp. 60–90.

33. Fox[7], p. 128, Fig. 78.

34. H.E. Maxwell, 'Notice of an Enamel Bronze Harness Mount from Auchendolly', *Proc. Soc. Antiq. Scot.*, XX (1885–6), p. 396.

35. H.N. Savory, 'A Find of Early Iron Age Metalwork from the Lesser Garth, Pentyrch, Glamorgan', *Arch. Camb.*, CXV (1966), p. 28, pl. II and Fig. 1.

36. Harrod[8], pl. XXXVII.

37. R.G. Collingwood, 'The Fosse', *Journ. Roman Studies*, XIV (1924), p. 256.

38. *Antiq. Journ.*, XVIII (1938), p. 40.

39. Maxwell[34], pl. VII:1.

40. *Proc. Soc. Antiq.*[2], XX (1903–4), pl. opp. p. 57.

41. C.N. Moore, 'Two Examples of late Celtic and Early Roman Metalwork, from South Lincolnshire', *Britannia*, IV (1973), p. 153, Fig. 1:a.

42. J.P. Bushe-Fox, 'Richborough IV', *Soc. Antiq. Research Report*, p. 106, pl. 1:2.

43. *Arch. Camb.*, CXV (1966), pp. 27–44.

44. J.B. Ward Perkins, 'Iron Age Metal Horses' Bits of the British Iron Age', *Proc. Preh. Soc.* (1939), pp. 173–92.

45. R. Munro, *Ancient Scottish Lake Dwellings* (1882), p. 132, Fig. 148.

46. V. Gordon Childe, *The Prehistory of Scotland* (Kegan Paul, 1935), p. 230, pl. XV.

47. Bulleid and St George Gray[25], p. 219, pl. XLII:E.159, E.174.

48. J.D. Cowen, 'A Cheek Ring of a Celtic Bit', *Proc. Soc. Ant. Newcastle*[4], VI (1933–4), p. 223, pl. XII.

49. Curle[19], p. 302, pl. LXXV:6.

50. Greenwell, 'Early Iron Age Burials in Yorkshire', p. 281, Fig. 23.

51. J.A. Smith, 'Notice . . . of a Massive Bronze Armlet . . . from Stanhope, Peeblesshire', *Proc. Soc. Antiq. Scot.*, XV (1880–1), p. 316, Fig. 3. For the patera, see Curle[12], p. 301. Probably it belongs to the early second century.

52. J.B. Ward Perkins, 'Two early Linch Pins from King's Langley, Herts., and from Tiddington', *Antiq. Journ.*, XX (1940), p. 358.

53. Ward Perkins[31], p. 65, pl. XI.

54. Smith[13], Fig. 4.

55. Brailsford[3], p. 222, and illustration.

56. Smith[13], pl. XV.

57. C.F.C. Hawkes and M.R. Hull, 'Camulodunum', *Soc. Antiq. Research Reports*, XIV, p. 331.

58. Fox[27], p. 43.

59. Hawkes and Hull[57], pl. XCIX:8.

60. R.A. Smith, 'Specimens from the Layton Collection in Brentford Public Library', *Arch.*, 69 (1920), p. 22, Fig. 22.

61. *BM Guide to Early Iron Age Antiquities*, Fig. 171.

62. R.E.M. Wheeler, 'Maiden Castle, Dorset', *Soc. Antiq. Research Reports*, XII, Fig. 90:6, and pl. XXIX:B:1.

63. Ibid., p. 274.

64. Smith[60], p. 22.

65. *V.C.H. Cambs.* I, p. 297, Fig. 28.

66. Smith[60], p. 22.

67. R.C. Hoare, 'Account of Antiquities found on Hamden Hill, with Fragments of British Chariots', *Arch.*, 21 (1827), p. 41, pl. VI.

68. Fox[27], pl. III:41.

69. Fox[7], pls. 16 and 17.

70. Wheeler[62], p. 275.

71. de Navarro in *The Heritage of Early Britain*.

72. Powell, *Préhistoric Art*, p. 204.

73. S. Piggott, *Early Celtic Art*, pp. 12, 49.

74. O.-H. Frey, 'Palmette and Circle: Early Celtic Art in Britain and its Continental Background', *Proc. Preh. Soc.*, 42 (1976), p. 55.

11. Mirrors

''Tis holy sport to be a little vain.' This was Luciana speaking, and presumably she was full sure of herself. Several Celtic mirrors have been found in the graves of women, and they provide useful indications of taste and trade in the luxury market in the first century of our era. No one knows who first thought of producing a refulgent surface, so that women could admire their reflections; but the probability is, so far as Britain is concerned, that the first mirror was an import.[1]

But imported mirrors were simple and fairly plain, whereas Celtic mirrors are highly decorated. A discussion on mirror decoration makes a suitable follow up to what has been said in the last chapter concerning patterns on horse trappings. Some elements in the designs are common to both. The whole concept of the design layout on the Polden Hill mount (Fig. 23:1) compares well with that on the reverse side of the Desborough mirror.[2] The obvious kinship here is further strengthened by the proliferation of S-patterns, which also occur four times on the Polden Hill mount. On the Desborough mirror, as on the Birdlip mirror, these S-patterns might appear to lack meaning, but the Polden Hill mount supplies the answer: the whole basis of the pattern is the cable motif. Here is an indication that the art of decoration had moved on quite a bit since the Polden Hill mount was made, and that in the process the significance of the pattern had been forgotten. The link between these patterns is strengthened again by the occurrence of basketry filling on another Polden Hill mount (Fig. 23:2). With the enamelled mount, full meaning has been given to the design by the secondary linear patterns. The switch from enamel-backed designs to purely linear patterns, such as those on the reverse sides of the mirrors, appears to have encouraged a steady devolution, which in the end brought about a breakdown in both quality and the exactitude of all patterns. It is easy to be careless or slipshod when making use of linear patterns, but the contrary is the case with champlevé enamel work.

Mirrors are assembled from three separate parts: a thin highly polished bronze disc, bounded by a rounded bronze edging, to give it strength, the whole being riveted on to a cast bronze handle. Sometimes a mount is placed in the bottom arc of the disc, between it and the handle. The reverse sides of these highly polished discs are usually elaborately decorated. The decoration can be as elaborate as that on the Desborough mirror, or as simple as that on the Rivenhall mirror. The patterns are always linear, because of the thinness of the metal. Handles are cast; and sometimes both they and the mount may be to some extent enamelled. Strangely, there is often a lack of harmony between the mirror decoration and the decorative moulding of the handles, the handles standing apart as though they belonged to a different tradition.

Cyril Fox ascribed mirror manufacture to the period between the last quarter of the first century BC and the time of the Claudian Conquest,[3] but these suggested dates need

109

emendation. The last in the series of mirrors is probably that from Nijmegen. It was found in the civil cemetery of Ulpia Noviomagus, along with cremated bones, and contained in a glass urn of late-first- to early-second-century type.[4] The Nijmegen mirror provides us with a *terminus ad quem* for the mirror style, which in this case is disjointed and devolved. On the other hand, Fox's *terminus a quo* date is far too early for the south-western school, to which Leeds[5] attributed some of the finest efforts of the geometrical phase of pattern-making, and to which the Birdlip and Desborough mirrors clearly belong. This school is more likely to have flourished in the first half of the first century AD, with a date of AD 1–15 for the Colchester mirror,[6] since it is assumed to have been one of the first made. Fox dated the Birdlip mirror to AD 25, although it had been found among burials ranging in date from BC 50 to AD 47.[7]

Fig. 30: Bronze Mirrors: Design on Reverse Sides 1. Stamford Hill, Plymouth. 2. Billericay, Essex. 3. Mirror in the Mayer Collection, Mersey Co. Museum. 4. Old Warden, Bedfordshire. 5. Colchester, Essex.

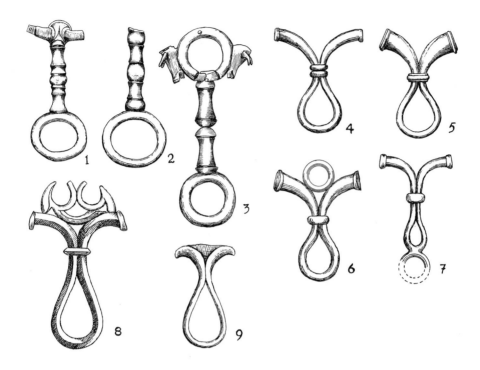

Fig. 31: Bronze Mirror Handles 1. Stamford Hill I, Plymouth. 2. Carlingwark, Kirkcudbright. 3. Ingleton, Yorkshire. 4. Trelan Bahow, St Keverne, Cornwall. 5. The Mayer Mirror. 6. Isle of Portland. 7. Mirror in Gibbs Collection. 8. Stamford Hill II, Plymouth. 9. Rivenhall, Essex. (Approx. 1/3)

The decoration on the Colchester mirror (Fig. 30:5) is unremarkable, being executed without true understanding; and included in the design are none of the patterns said to be typical of the more advanced decoration of the south-western school. Other mirrors, from eastern England, such as those from the Essex area and from Old Warden in Bedfordshire (Fig. 30:4), also show the same lack of understanding. This inferior knowledge of Celtic designs in eastern England is understandable when the form of the mirror handles is taken into consideration. The same region of eastern England saw numerous occurrences of Roman mirrors,[8] with the largest number coming from the Colchester cemeteries. Of particular interest to us is the handle-form, which is either baluster shaped or in the form of a loop-shaped grip, the single-loop form being the most common (Fig. 31). By turning now to the Celtic mirrors from the same region, it will be seen that the looped handles are also the most common, but that baluster-shaped handles also occur, though in much smaller numbers. Thus, the study of the home-produced mirrors now takes on an added interest, in that clearly their handle-forms have been copied from the Roman model, which got to Britain by the normal trade routes, long before the Claudian Invasion took place.

South of the Jurassic Ridge there appear to have been two situations: on the one hand there were native copies of the Roman form, occurring mostly in the Essex area; whilst on the other hand there were the products of the south-western workshops, which appear to have received their inspiration from St Mawgan and Polden Hill. The Desborough mirror is kidney shaped, as are the mirrors from Billericay and Old Warden. The form of the Old Warden handle has been inspired by the shapes of the handles from Desborough and Birdlip. But the basis of the design of these handles (Fig. 7:1 and 2) is still the loop, but here utilised in multiple form by being conjoined back to back. The appearance of both baluster and looped handles side by side at Stamford Hill[9] must indicate a certain amount of contact with the eastern area, or perhaps with Provincial Roman sources through trade channels; but the mirror itself is firmly in the western tradition, as its linear patterns demonstrate; whilst an aura of respectability has been provided by an addition to the top of the handle of an underlying Celtic motif.

A very Celtic motif, the spherical triangle, was incorporated into the linear patterns on the western mirrors, though actually it did not form part of any one pattern. Instead, enclosed within a circle, it was almost hidden somewhere within that pattern. Only one eastern mirror, the Colchester mirror, possesses this motif.

Fox looked upon the Mayer mirror (Fig. 30:3) as being the earliest. Rightly, he speaks of the perfection of its linear pattern on the reverse side; but this mirror cannot be as early as he imagined. The design is made up of three roundels, each linked to the other, and each containing decoration that is entirely complete in itself. This decoration is of a very pure Celtic character, superbly carried out, both artistically and technically, and in development it presages the eventual separation of the roundels, as has happened in the case of the Trelan Bahow mirror.[10] The Mayer decoration consists of a curious combination of non-boss-and-petal motifs or patterns, developed out of the cable-motif based S-scroll of the Polden Hill mount: or, alternatively, it might represent a development of the pattern seen on the bronze plaque from Knobb's Crook (Fig. 26:4). In either case, development has come some way, and therefore the Mayer mirror must be later than either the Birdlip or the Desborough mirror; for in their cases the linear designs on the reverse sides incorporate several fairly well represented lotus-bud motifs, with tendrils swelling out into volutes which are interrupted by the presence of lobes of similar form to those seen on the harness mounts. This gives to these two mirrors a closer relationship to the harness mounts and their own forms of decoration.

In retrospect, the patterns on the Birdlip, Desborough, Mayer, Trelan Bahow and Stamford Hill mirrors must have a common origin in a western school; for, even culturally, as Leeds has pointed out, Northampton belonged, or was closely allied to the western school. The Trelan Bahow mirror must be of early Occupation date, with its edging of small hatched triangles, suggesting an analogy with others on one of the Polden Hill mounts (Fig. 24:3) and the Westhall terret (Fig. 27:3), two objects which are most likely to date from the middle of the first century.

In contrast to the decoration discussed above, the patterns on the Billericay,[11] Old Warden,[12] and Rivenhall[13] mirrors belong to an art in rapid decline. The wild abandonment of principles is in marked contrast to the controlled layouts of the patterns on the Birdlip and Desborough mirrors. Yet, amongst so much confusion, the Old Warden handle stands out as a silent reminder of contact with the west.

The decorative pattern on the Old Warden mirror (Fig. 30:4) resembles nothing so much as a badly conceived jigsaw puzzle; but on the one from Colchester there is a more formal pattern, a simpler version of what might be described as the western style. There are underlying motifs that call to mind a reversed form of the Brunn-am-Steinfeld lotus-bud, as seen on the Polden Hill mount[14] and the Desborough mirror. If the Old Warden handle can be taken as our guide to date, then the mirror must come within the early Occupation era. The explanation for that statement can be seen in the design on the upper part of the handle, where it has been extended to form a mount. This design calls to mind the breaking up of the swash-N pattern, noted on objects such as the bronze buckle from Caerleon, Monmouth (Fig. 56:3) and on another like it from Richborough, Kent.[15] Both of these should date after AD 75. Thus, mirrors of the Old Warden style must therefore belong to the fourth quarter of the first century; and this must be our guide to the dating of most of the Essex mirrors, especially of those like Billericay I, and the 'Disney' mirror, which, so far as decoration is concerned, bear some kinship to one another. But, this line of argument still leaves unanswered all questions concerning the date of the Colchester mirror, on which the broken-off ends of the swash Ns (witness the Old Warden mirror) have here got transferred to the linear pattern on the reverse side. The handle is a little in keeping with the Old Warden one; yet with all this, the mirror accompanied a rich cremation burial which included Belgic pottery of exceptional quality, and a spun bronze handled cup. Drawing on his detailed knowledge of early Colchester, the late M.R. Hull reckoned that this pottery could have been made during the reign of Cunobelin, with a suggested date range for it of AD 10–25.[16]. If the Old Warden style of decoration means anything at all then this pottery date is half a century too early for the mirror; so that the anomaly must remain.

Although the whole basis of mirror-style decoration is the swinging curve, recognisable motifs and patterns make their appearances. Sometimes there is confusion with regard to the use of basketry filling. On the Colchester mirror it is the pattern, and not the background, which is basketry filled: on the Birdlip and Desborough mirrors the reverse is the case. Confusion of thought is apparent in the case of the Stamford Hill mirror[17] (Fig. 30:1), for both filled parts and plain parts have messages for us, and one wonders to which was given the priority. In the plain part of that area included within the roundels are some credible representations of the lotus-bud motif, the tendrils being drawn out to merge with the circle of the roundel. On looking at the decoration, one realises that both St Mawgan and Polden Hill are not very far away, and influences from both are evident here. In fact there is a more intimate knowledge here of original and recognisable motifs, as compared with instances elsewhere, where, as Fox has stated, a more studied formlessness exists.

As a centre piece of its decoration, the Billericay I mirror has a 'twin circles within a third circle' design, from which radiate the three arms of a triskele (Fig. 30:2). This little central pattern is here very stylised: in less stylised form it occurs on the Thames at Datchet spearhead (Fig. 32:4). Whereas on the Billericary mirror the circumferential line is continuous, on the spearhead this line is not truly continuous, but is on a slightly reducing radius, meaning that the circuit cannot be complete, thereby leaving a slight gap. In this representation there is a clue to the origin of the pattern; for it is derived from similar patterns at the centres of the right-hand and left-hand roundels on the reverse side of the Stamford Hill mirror. How would one find a name that would cover this rather unsatisfactory description; but the pattern is easily recognisable for all that. As we see it on the Stamford Hill mirror, so does it occur in running form on one of the sword scabbards from Lisnacrogher (Fig. 36:1). But the Irish added dots. The similarity of other Irish patterns to those in Britain is the subject of past and future comment: these extend to basketry filling and lotus-bud representations.

Workmanship at Stamford Hill was not up to the usual western standard: exactitude is lacking in the representation of the overall pattern, which may be due to incoming decadence which is reflected in the use of a baluster handle. But the design which started this discussion, namely that at the centre of the Billericay mirror, continued in use elsewhere, at Meare[18] (Fig. 32:2) and again in the second century at Milking Gap, Northumberland,[19] appearing there on a dragonesque brooch (Fig. 53:4) which was lost *c*. AD 122.

Some of the western mirrors were found in graves containing inhumations. The Trelan Bahow mirror was found in 1833 along with bronze bracelets and rings, of uncertain use, and two glass beads, in one of several interments protected by covering stones and slabs set on edge. At Birdlip three bodies had been interred in a continuous line: the middle grave contained a female skeleton, of a woman in the prime of life, and the mirror had been buried with her. Whitewashed slabs protected this grave. The Stamford Hill mirror was found in association with brooches similar to some Roman types, for which a date in the second half of the first century is apposite. These extended inhumations were those of a dolicho-cephalic people, with an index of 76 for Birdlip. In his study of the Glastonbury skulls, Boyd Dawkins[20] concluded that the people of the lake village possessed a strong affinity to Mediterranean race. This line of enquiry is too often ignored when assessing relationships in art, yet clearly race must have some bearing on artistic expression. A strong distinction can be drawn between the funerary practices here in the south-west and those in Essex, where cremation was the rule. The Essex rite was under strong Provincial Roman and Belgic pressure; for, with the Belgae, cremation was also the normal burial rite. British princes and nobles, many of whom must have been of Belgic descent, took the lead in developing Romanising tastes, and, as Hawkes states, the spread of Romanisation accompanied Augustus' prosecution of diplomatic relations with the British.[21]

Clearly, therefore, there was a division in cultural traditions between eastern England and the older and more traditional south-west; and this, in no small measure, must have been due to differences of race. But cultural contacts must have existed, as we see with reference to the mirror handles, which so often are out of harmony with the decoration on the discs. There can be but one explanation for this: the Celts had no handle design of their own, so they borrowed a pattern from elsewhere. Only three handles are consciously Celtic, which are those belonging to the Birdlip, Desborough (Fig. 7:1 and 2) and Old Warden mirrors; and even here they are made up of simple loop forms borrowed initially from Roman or Belgic sources. But these three handles must represent genuine attempts to arrive at a design which would harmonise well with the linear decoration, and in this respect the craftsmen were successful. A selection of simple Celtic-made handles based on the Roman pattern is shown in Fig. 31. Here are two styles: the baluster and the loop. The latter is by far the more common, and it must have appealed to native craftsmen because of its curves. There are only three reported discoveries of baluster handles: from Stamford Hill, from Ingleton, and from Carlingwark, and only the first accompanied a mirror. But the looped handle is widespread. In British hands, the cast-in collar at the centre, where the loop crosses, clearly indicates a permanent substitute for a simple tie. The otherwise simple ends of the loop now swell into trumpet forms, with beaded-edge mouths. This tendency has been carried to a marked extent in the case of the Mayer mirror handle. The same trumpet form was utilised for the S-brooches of a decade or two later. That so simple a handle should have been used for the Mayer mirror may have resulted from design rather than lassitude: for, if by design, then the intention was that nothing should be permitted to attract the eye away from an uninterrupted contemplation of the linear patterns. At Uplyme, Devon,[22] the treatment was quite different. Here on the Holcombe mirror the mount is decorated with double trumpets (Fig. 15:3), and carried out in typical Celtic style. One gets the impression that the native feeling for line was a prerogative of south-western artificers, and that this trait is manifested in their treatment of the simple loop-form of handle. In the case of Fig. 31:6, from the Isle of Portland, a simple bronze circle has been added; but the Stamford Hill maker was more venturesome, in that he added openwork decoration which included within its form the lotus-bud motif. Clearly, two cultural drifts were on a collision course in south-west England, but the results perhaps were of benefit to both.

Double-trumpet decoration is not confined to the Holcombe mirror; it is also part of the design on the Birdlip mirror mount, on which it occurs in addition to that nameless pattern consisting of two circles within a third, a devolved form of the pattern on the Stamford Hill mirror, the Lisnacrogher scabbard and the Thames at Datchet spearhead. These western products help to highlight the survival of the later pre-Roman Iron Age civilisation, which continued to receive stimulating art motifs from Europe at a time when the east coast of England was receiving, through the normal trade channels, Provincial Roman objects from Gaul.

North of the Jurassic Ridge conditions were quite different. Here iron mirrors tended to take the place of bronze: one was found at Garton Slack,[23] buried with a female skeleton. The handle had mountings of bronze. Other iron mirrors have been found at Arras; and the Ingleton handle had been meant for an iron mirror. The only mirror handle so far found in Ireland may likewise have been intended for an iron mirror (Fig. 33:3).

Finally, it must be said that these mirrors have provided a fascinating insight into the luxury market of the first century AD. Mirrors must have been expensive objects in their day, particularly specimens like the Desborough and Birdlip mirrors. The women who commissioned them from the King's craftsmen must have been both vainglorious and well endowed. Yet the wives of the rich Wessex farmers never bothered with them, and there is no evidence that the beaux of those days got pleasure from their reflections. For, who would set his wit to so foolish an exercise as the study of his profile? But then, as Bottom said, man is but an ass.

NOTES

1. G. Lloyd-Morgan, 'Roman mirrors in Britain', *Current Archaeology*, 58 (1977), p. 329.

2. R.A. Smith, 'On a late Celtic Mirror found at Desborough, Northamptonshire, and other Mirrors of the Period', *Arch.*, LXI (1909), p. 329.

3. Fox, *Pattern and Purpose*, p. 84.

4. Dunning, 'An Engraved Bronze Mirror from Nijmegen, Holland', p. 69, Fig. 1.

5. E.T. Leeds, *Celtic Ornament* (Oxford, 1933), p. 32.

6. The Colchester Mirror was found with late Belgic pottery, which M.R. Hull (*Antiq. Journ.*, XXVIII (1948), p. 136) suggests was made in the reign of Cunobelin (AD 10–43) with an amended date for the pottery of AD 10–25.

7. C. Green, 'The Birdlip Early Iron Age Burials: a Review', *Proc. Preh. Soc.*, XV (1949), p. 188.

8. Lloyd-Morgan[1], p. 331.

9. C. Spence Bate, 'On the Discovery of a Romano-British Cemetery near Plymouth', *Arch.*, XL (1866), pl. XXX:1.

10. *BM Guide to Early Iron Age Antiquities*, p. 121, Fig. 132.

11. Smith[2], p. 337.

12. Ibid., p. 333, Fig. 3.

13. Ibid., p. 337, Fig. 5.

14. Fox[3], p. 87, Fig. 52, for illustrated details of similarities.

15. J.P. Bushe-Fox, 'Richborough IV', p. 123, pl. XXXIII:73.

16. Hull[6], p. 136.

17. Spence Bate[9], pl. XXX:1.

18. *Antiq. Journ.*, X (1930), p. 154.

19. Kilbride-Jones, 'The Excavation of a Native Settlement at Milking Gap, High Shield, Northumberland', p. 342, Fig. 5.

20. Bulleid and St George Gray, *The Glastonbury Lake Village*, II (1915), pp. 673ff.

21. T.D. Kendrick and C.F.C. Hawkes, *Archaeology In England and Wales* (Methuen, 1932), p. 204.

22. A. Fox and S. Pollard, 'A decorated Bronze Mirror from an Iron Age Settlement at Holcombe, near Uplyme, Devon', *Antiq. Journ.*, LIII (1973), p. 29, Fig. 8.

23. T.C.M. Brewster, 'Garton Slack', *Current Archaeology*, 51, p. 109 and illustration.

12. Weapons

Immediately, one thinks of swords and spears. British early-Iron Age swords and scabbards have been dealt with comprehensively by Piggott.[1] Naturally, this extensive coverage is rather more than is required here, since we are dealing with the final half century of manufacture before sword-making amongst the British in Britain became a proscribed occupation. And, in a work of this nature, it is not so much the swords themselves which are of immediate interest, but their scabbards, often made of bronze, which bear interesting and sometimes beautiful Celtic-type decoration. Some of this decoration is in the mirror style: there are the usual flowing curves and suggested motifs, all combining to make up patterns which call for analysis. Decoration on the British sword scabbards is more often confined to the upper parts, where it can be more easily seen; only in Ireland is decoration continuous over the whole length of the scabbard.

The swords are narrow-bladed, tanged, and equipped with arched hilt guards. The ones chosen belong to Piggott's Group II. The distribution stretches from Somerset northeastwards up the Jurassic Ridge to Hunsbury, and thence to the east Riding of Yorkshire. Some are gathered together in the Cambridgeshire Fens, but the main concentration is in the Thames Valley. The scabbards are made both of bronze and of iron, but the iron scabbards are less well preserved. Three of the bronze scabbards have finely engraved or chased ornamentation upon them. These scabbards are made of two bronze plates, the outer one, which carries the decoration, being folded over at the edges, in order to hold the back plate in position. A main feature is the chape, which is a separate bronze casting, fitted to the pointed end of the scabbard, in order to give it strength and protection. Scabbard chapes go through an evolutionary process, but this does not concern us here. Finally, on the reverse sides of the scabbards there is a metal loop, for attachment to a belt.

Of the four scabbards illustrated, clearly the earliest is Fig. 32:1, which comes from Hunsbury, Northamptonshire.[2] Associated finds ranged from la Tène I brooches to Belgic pottery, but the relationship of the scabbard to any of these finds is not known. This specimen belongs to Piggott's Group II. The only specimen of this Group with any claim to date is one from Bredon Hill, Gloucestershire,[3] which was found in a first-century AD context, thus indicating manufacture of at least some swords of Group II in the pre-Invasion period covered by the present work. The Hunsbury scabbard could well belong to the same period, for the decoration consists of interconnected roundels, each containing circles and non-boss-and-petal patterns, together with simulated double-cable motifs and basketry filling — all very reminiscent of the early mirror style. In fact, the decoration in the Hunsbury roundels compares well with that in the roundels of the Mayer mirror. The style is precisely the same. On the reverse side of the scabbard, and about

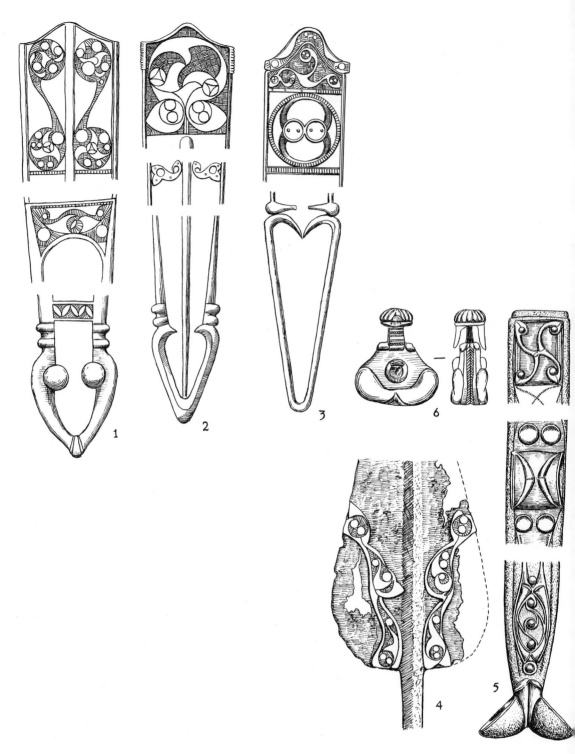

Fig. 32: Group II Decorated Sword Scabbards 1. Hunsbury, Northamptonshire. 2. Meare, Somerset. 3. Amerden, Bucks. Group IV Decorated Sword Scabbards 4. Decorated Spearhead, Thames at Datchet. 5. Mortonhall, Edinburgh. 6. Sword Pommel, Brough-under-Stainmore, Cumbria. (1/2)

half-way down its length, is a hatched-out linear pattern, which is very similar to others seen in open metalwork at Stanwick. Such patterns were derived from more elaborate examples, like those on the Santon harness mount (Fig. 24:1). In their way, they are versions of the broken-back scroll, the shape of which has been influenced by the cable motif, which may be seen in numbers on the Santon mount. As is well known, many hoards were buried about the middle of the first century. Therefore, there is a strong possibility that the Hunsbury scabbard belongs to the first half of that century.

A scabbard which, from its style of decoration, must be somewhat later is that from Meare, Somerset,[4] which was a chance find somewhere near the site of the lake village. The decoration (Fig. 32:2) is much nearer to the mirror style, and it looks like the work of the south-western school, discussed in the last chapter. The style is still very much in the same tradition as that represented on the Mayer and Trelan Bahow mirrors, and there is a remote possibility that all may have come from the same workshop. The apparent odd imbalance in the overall design is perhaps one of its chief attractions, and typical also of both the Mayer and the Trelan Bahow patterns. The basketry filling is also on a par with that on the Mayer mirror. In all, the non-boss-and-petal pattern is very prominent, whilst in the case of the Meare scabbard, there is also a link with Stamford Hill, in the similarities of the nameless pattern of two circles within a third, again discussed in the last chapter. Here it resembles its manifestation on the spearhead from the Thames at Datchet,[5] except that on the spearhead this pattern is becoming abstract. The spearhead (Fig. 32:4) is of iron, and the curious and eccentrically shaped bronze plates, attached to it by means of pins, may have had their shapes determined by the layout of the decoration. Plates occur on both sides of the spearhead, and no two are alike, either in form or in the matter of the curvilinear patterns incised upon them. Perhaps these patterns are a little less elegant than are those upon the better mirrors, but the restriction in space may have provoked a tendency towards convention. However, the same basketry background is there, but the general style is a little earlier than Meare, and it is perhaps on a par with Hunsbury.

How common were these patterns? It would seem that they were reserved for flat surfaces on thin metal, on which they are carried out in linear style with the help of a chaser. Apart from such oddities as the crescentic strip from Balmaclellan,[6] on which there are basketry-filled continuous S-scrolls and a swash N on a wedge-shaped panel, the styles we have been considering, like the mirror style itself, are unreservedly southern. Since enamel, if it is seen at all, plays a very small part in the story, Belgic influence must be very slight, or absent altogether: and it is also possible that this decoration style came into being to rival the enamel work of the Belgae; perhaps as a reaction on the part of the later pre-Roman Iron Age peoples to those Belgic enamelled designs. Of course, in the end, enamelling won the day, as it was certain to do, because it is attractive, even at a distance. All would be part of that east/west division of styles of decoration, noted in the case of the mirrors; for it is worth pointing out in passing that, in spite of the known excellence of their metalwork, the Iceni never made use of the mirror style of decoration.

That this style, which is common to scabbards and mirrors alike, managed to exist until the last quarter of the first century is proved by the existence of the Nijmegen mirror. But on it the patterns have broken up, appearing without any real linkage. The curious handle, or what remains of it, is decorated at the centre with a rosette, which is partly enclosed by raised quadrants (Fig. 21:2); whilst the rest of the handle has crescents of enamel interrupted at their middles by dots enclosed by metal. It is a difficult design to describe, chiefly because of its unusual character: but another, very similar, can be found on a scabbard from Amerden, Buckinghamshire[7] (Fig. 32:3). However, here the central rosette has been replaced by two circular recesses, formerly occupied by buttons, since their rivet holes are still visible. Once again, a design like this must owe something to that design of two circles within a third seen on the Stamford Hill mirror, the Thames spearhead, and other examples quoted above; and here at Amerden the decoration is perhaps half-way between Meare and Nijmegen. And above this pattern on the Amerden scabbard there is a triangular panel within which there is a three-way scroll, perhaps meant for a triskele; and the arm ends are swollen into non-boss-and-petal patterns, the places of the non-bosses being taken by this curious pattern which we now know so well, having familiarised ourselves with it as it occurs on the Stamford Hill mirror, the Thames spearhead, and the Lisnacrogher scabbard. It persists in cropping up at this time, and basically it is a western pattern, for in the west it occurs in unstylised form. Basketry filling on the Amerden panel bears out the same thought.

Leaving aside the Group II scabbards, the picture changes dramatically when we come to a consideration of Piggott's Group IV scabbards, which are Brigantian. Piggott believes that swords and scabbards of this Group fall within the period *c*. AD 45–125. The distribution is entirely Brigantian, except for one stray found at Mortonhall, Edinburgh.[8] A small but curious concentration occurs in Dorset, and the swords here are unlikely to post-date the Invasion. Some of the Stanwick swords are Flavian; but most, if not all must have been made before the disaster of AD 74.

Decoration is at a minimum on these Group IV scabbards, and what there is of it is usually lavished on the centrally placed strap-loops; for these scabbards were suspended from their middles, with the loop facing outwards, in contradistinction to the scabbards of Group II. A continuous applied metal strip connects ornamented terminations top and bottom. At Mortonhall (Fig. 32:5) a tendency towards the bizarre is suggested by the bifid chape; but its grotesqueness is relieved by the finely executed decoration at the bottom end of the central strip; decoration which is in the finest tradition of Celtic art, and it features the broken-back scroll. As with that at Lochar Moss, the Mortonhall style is a semi-boss style; and the broken-back scrolls are enclosed by two pairs of trumpets identified as being in the Irish tradition. In the form in which they are represented at Mortonhall they have parallels on the Navan brooch (Fig. 35:3), on which they enclose a small roundel intended for some insert, like a button. Other trumpets, not in the Irish but in a north British tradition can be seen cast into the chape. There is a marked distinction here between the two forms of representation: the chape trumpets are more like those on the Holcombe mirror, and they are of a type current just before 'capping' took place.[9] The decoration at the top of the central strip is almost swastika-like in layout, but in reality the four arms are not conjoined at the centre, as they appear to be: instead they are merely in touch. The arms have been taken straight out of the swash Ns of the Plunton Castle armlet (Fig. 15:2), thus confirming what was hitherto suspected, that the trumpet-enclosed broken-back scrolls at the bottom end of the loop strip are carried out in the Galloway style.

What is the sum total of this evidence? Piggott[10] speaks of the utilisation of bronze alloys in contrasting tones on the Mortonhall scabbard, to emphasise (as he says) the openwork patterns of the applied loop-strip. This may or may not have been the craftworker's intention. Suppose Piggott's theory to be in doubt, then the probability is that the Mortonhall scabbard was brought north by somebody fleeing from the Romans, and that, being in a different environment, he had the loop-strip replaced by one made locally. In other words, this strip with its Galloway-style decoration represents a repair, and this would explain the difference in alloy colours. Such a decorated strip could never have been applied in Brigantia since the decorative style was not current in that region. Otherwise, the Mortonhall chape is matched by the chape on the scabbard from Cotterdale, north Riding of Yorkshire;[11] and, if anything, that parallel serves to draw attention to the very ordinary plainness of what passes for decoration in Brigantia. Piggott saw in the Mortonhall scabbard a non-functional parade piece. If it had been brought north following the disaster of AD 74 there was still time to have it repaired by Galloway craftsmen in the Galloway style, before Agricola stationed his men all along the western coasts of Galloway.

The suppression of Celtic tribal authority by the Romans robbed the native sword-makers of their patrons. Some of these unemployed sword-makers sought jobs in the Imperial workshops. An example of their work is seen in the sword possessing a modified form of the cocked-hat hilt guard of Piggott's Group IVa, which came from Fendoch, Perthshire.[12] A form of pommel that could have been produced under these conditions is that shown in Fig. 32:6, which comes from Brough-under-Stainmore, Cumbria.[13] It has been included in Piggott's Group IV possibly because it was found in Brigantia. But it has a fluted cap, a central boss, and two linked marginal lobes by way of decoration; whilst chevron pattern decorates the sides. It is matched by only one other, the sword-hilt from Worton, Lancashire.[14] The marginal lobe decoration is reminiscent of that on the linchpin from between Middleton and Enthorpe in the east Riding of Yorkshire (Fig. 28:2) though there are essential differences which could put these lobes beyond any easy derivation. The chevron pattern suggests Provincial Roman influence, confirmed by the fluting on the cap. There is very little more that can be added, except to say that in all probability both came from the same workshop, which may have been at Brough itself.

The small number of decorated scabbards extant must be taken as a degree of the success of Roman measures taken during the early days of the Occupation, when all arms had to be handed in.

Shields

It is appropriate that shields should be discussed in the present chapter, since swordsmen invariably carried them on their left arms. But shields of our period are not plentiful, some of the best examples being earlier; but there is one which is firmly within our period, and this is the superb specimen which was found in the River Thames at Battersea, London.[15] Preserved is the decorated bronze facing for a backing of wood, the two having been secured to one another by U-sectioned binding. Decoration is confined to three roundels: a central one connected by decorated extensions to slightly smaller terminal roundels, numbering two. All these are in three separate pieces, secured to the backing by rivets (Fig. 1). The overall length of the shield is 85 cm.

Much has been written about this shield, and what has been remarked upon is the mechanical quality of precision which attended the laying out of the decoration. Of course, this is all good compass work, comparable with the precise layouts of broken-back scrolls seen on the trial pieces from Loughcrew, where part of a compass was found. Technical mastery is therefore evident on both sides of the Irish Sea in the first half of the first century AD. The decoration on the Battersea shield is carried out in repoussé style, but here it is very restrained, and there is none of the flamboyance suggested by the pattern on the Torrs pony-cap. Instead, the work has been carefully controlled, so that now the punched-up lines of the pattern are already suggesting the spidery style seen on the Cork horns, the Petrie Crown and the Bann disc, discussed in the following chapter. Close contact between bronzesmiths on both sides of the Irish Sea is further attested by the similarity of the decoration on the Cork horns with that on the Stichill collar.

Back now with the Battersea shield, it will be seen that the general layout of roundels and their contained decoration is along the lines suggested by the Witham shield, though it must be clear that the art has come a long way since those earlier times, not only with regard to the decorative style, but with regard to technical ability as well. The one decorative feature that everyone refers to is the much simplified form of the palmette, which is to be found amongst the decoration on the central roundel, and of which there are two examples. In these palmettes, the spiral ends have given way to circular glass inserts, each relieved by a swastika pattern. Anyone curious as to why these spiral ends were dropped in preference for circular glass inserts will find the answer by looking carefully at the composition of the palmettes, and he will see that each is made up of two cable motifs, a pointer to the existence of which is the ogee-shaped central punched up line. In so far as double-motif motivated patterns are concerned, it will be remembered that the coupling together of motifs was already evident amongst the decoration on the Aesica brooch; but that at Aesica the idea was tentative, whereas here at Battersea the mixing has been done with a sure hand. Such confidence implies that the Battersea shield is later than is the Aesica brooch. It is clear that the conservative approach to motifs has been abandoned in preference for a certain fluidity in this matter of the incorporation of more than one age-old motif in the patterns of the day; and this achievement, as far as the palmettes are concerned, has put them but a step away from the simplified versions seen on the horn-cap recovered from the River Thames at Brentford (Fig. 29:2); whilst the swastika embedded in the round-glass inserts can be paralleled at Llyn Cerrig Bach on another horn-cap (Fig. 29:1), a typologically early specimen. A typologically later specimen from Maiden Castle[16] has been dated by Wheeler to the period AD 25–45.

Still on the subject of motifs, apart from the easily recognised palmettes, there are also examples of the lotus-bud motif. These occur in opposition on both terminal roundels. Altogether, there are two pairs to each roundel, and, as if to emphasise their importance, they alone are stippled. In form, these lotus-bud motifs are similar to those on the Polden Hill mount (Fig. 23:1) on which the representation is formal and very close to the classical tradition. The tendril-like ends have been bent to form curves in the Battersea decoration; but all is restrained, and there is nothing in the entire decoration to suggest the wild exuberance noted in the case of the decoration on the St Mawgan shield mount. Restrained styles are, of course, later. The rest of the decoration is made up of S and C curves; but, top and bottom on the central roundel, there are emergent raised quadrants. The connecting patterns (that is, those placed between the central and terminal roundels) are suggestive of the broken-back scroll based on the cable motif (already debated in the chapter on Trappings) most readily compared with the style of those from Stanwick (Fig. 24:5) and Eckford (Fig. 26:7).

We are now in a position to say a little more about the date of the Battersea shield. Whilst Reginald Smith thought of it as being later than the Wandsworth shield, which he judged on style, nevertheless the decoration suggests a mid-first-century-AD date. As pointers there are the swastikas, paralleled at Llyn Cerrig Bach; the formal lotus-bud motifs, similar to those on the Polden Hill mount; the broken-back scrolls based on the cable motif, and not unlike those at Stanwick – all these point, by their similarities, to others known to belong to round about the mid first century AD. In all probability, the Battersea shield was a votive offering, its splendid condition suggesting that it was never put to use in anger.

NOTES

1. Piggott, 'Swords and Scabbards of the British Early Iron Age', p. 1.

2. *Ass. Arch. Soc. Reports*, XVIII, p. 58, pl. III:3; *V.C.H. Northants*, I, pl. opp. p. 147.

3. *Arch. Journ.*, XCV (1938), pl. VII:2, opp. p. 13.

4. *Antiq. Journ.*, X (1930), p. 154.

5. T.D. Kendrick in *MAN* (Royal Anthropological Institute, 1931), p. 182.

6. A good drawing can be found in Macgregor, *Early Celtic Art in North Britain*, II, no. 342.

7. *V.C.H. Bucks.*, I, p. 186.

8. Kemble, *Horae Ferales* (1863), pl. XVIII:5; Macgregor[6], no. 150.

9. A typical example of the 'capped' trumpet can be seen on the belt-plate from Drumashie (this book, Fig. 15:6).

10. Piggott, *Early Celtic Art*, no. 32.

11. A.W. Franks, 'Notes on a Sword found in Cotterdale, Yorks.', *Arch.*, XLV (1880), p. 251.

12. I.A. Richmond and J. MacIntyre, 'The Agricolan Fort at Fendoch', *Proc. Soc. Antiq. Scot.*, LXXIII (1938–9), p. 146, pl. LX:1.

13. J.D. Cowen, 'A Celtic Sword Pommel at Tullie House', *Trans. Cumb. & West. Antiq. & Arch. Soc.*, XXXVII (1937), p. 67.

14. *BM Guide to Early Iron Age Antiquities*, p. 108, Fig. 117.

15. Kemble[8], pl. XV, p. 190; *BM Guide*[14], p. 106 and frontispiece; Fox, *Pattern and Purpose*, p. 27, pls. 14a, 16–17; Megaw, *Art of the European Iron Age*, no. 253; Brailsford, *Early Celtic Masterpieces from Britain*, p. 25, pls. 24–35, and pl. X.

16. Wheeler, 'Maiden Castle, Dorset', Fig. 90:6.

13. Craftsmanship in Ireland (I)

Irish metalwork never gets the publicity it deserves. The cause may be due to its highly individual nature, differing as it does in certain subtle ways from the British equivalent, and developing along different lines from those followed by Celtic art in Britain. Not very much now remains, which again encourages neglect. The paucity might have been surprising but for Wakeman's statement,[1] written down in 1889:

> It should be observed that until about thirty years ago very little attention was paid to the character of antiquities occurring in this country [Ireland], and that tons [*sic*] of early bronzes, consisting of arms, implements, and objects of personal decoration, dug up from our bogs, or dredged from river courses or lochs in all parts of Ireland, were consigned to the furnaces of native or British brassfounders, where they were simply melted down as old metal.

Therefore Irish studies will always face a disadvantage, in that many of the finest pieces are presumed lost, and many found today are suffering the same fate of being melted down. The purpose of this chapter is to take stock of what is left; and strangely the signs are more heartening than otherwise might have been expected. We can still get a fair picture of Irish metalwork. But the absence of trappings must be a cause of surprise. For the Irish, like other Celtic peoples, had the 'carpat', which can be translated as 'chariot': but it was not a war chariot.[2] 'It was in the *carpat* that a warrior drove to his formalised combats, and in it he performed the ceremonies of insult and display that preceded these combats.'[3] Also, equitation was known, evidence for which is there in the hoards and single discoveries of bridle-bits.[4] But these bridle-bits occur only in a narrow belt of country from Galway Bay to Dundalk Bay, and thence north, taking in Lough Neagh and the Bann valley. It is most important to remember that all the bronze objects about to be considered also came from the same areas as did the bridle-bits. To prevent confusion, this coincidence does not mean that all the objects are contemporaneous. But, because finds are distributed in this way, the rest of Ireland appears to have been a cultural desert.

Personal ostentation was much in evidence. Objects like the Cork Horns and the Petrie 'Crown' must indeed have been intended for the highly privileged. Irish art objects are noted for their flamboyance; but their period of manufacture remains open to some speculation. Whilst in Britain the Roman presence was the means by which a relatively accurate historical record exists, in Ireland there was nothing comparable. Even Irish decoration never compares closely with the British; but certain elements in it can be found to have parallels in Britain. But a further complication is the existence of a time-lag.

All manners of problems arise because of the introduction of an early style, the Waldalgesheim; and because the repoussé style achieved a popularity in Ireland which it never attained in Britain. Had it done so, the style would have been different. Also, why did a linear style leave Armorica and travel up the Irish Sea to Co. Antrim, touching in at St Mawgan en route? A good deal in Irish archaeology does not make for good sense; but we have to live with this situation, which is still very much before us. From the bold repoussé style, the Irish craftsmen created another, spidery in its delicacy, the technique of casting objects and patterns together as one;[5] and, in this respect, Irish metalworkers were ahead of all other bronzesmiths. The useful die-work in the Galloway style, noted on the Plunton Castle armlet, also should not be forgotten.

The story of early contacts between Ireland and the la Tène art world are beyond the scope of this book; yet they must be alluded to, since they are part of the picture we are trying to create here. So we report contact with continental sources at an early period, when the buffer-ended, Waldalgesheim-style gold torc somehow found its way to Clonmacnoise, on the River Shannon.[6] It might seem extraordinary that the only genuine piece in the Waldalgesheim style, ever found in the British Isles, should have turned up in such a remote area. But the matter is of considerable significance, because the area is within the lands inhabited by a people called the Soghain,[7] who also occupied Co. Meath; and it is from within 'the old plain of the Soghain' that the majority of art works have come, starting with the Turoe stone.[8] The art represented on this mid-first-century pillar stone is of a particularly pure nature, still continental in origin; and but for this record in stone, we should never have been aware of the existence of so sophisticated an art form at such an early period in Ireland. There is not a single metal object known which has decoration at this stage of development; which brings us back to Wakeman's statement. Even so, all is not lost; for the form of rounded relief given to motifs as represented on the Turoe stone was reproduced in bronze in the form of punched-up-from-behind patterns, here referred to as being in the repoussé style; and some of the earliest and best occur outside Ireland. Repoussé work is a laborious undertaking, since, in order to keep punching up the design, frequent heating of the metal is necessary, to prevent its becoming brittle, and to stop cracking. Overstretching of the metal, due to excessive punching, is frequent; the effect is seen in the parting of the metal along highly raised ridges. Clearly, there was a need for a substitute technique, which is why, in later years, repoussé was dropped for the technique of casting objects and patterns as one, and together. From the purely aesthetic point of view, the result was less satisfactory, since the patterns lost their bold appearance, and instead became sinewy. In this way decoration lost a great deal of impact. Linear patterns never seem to have been favoured very much in the lands of the Soghain, but were better thought of in the Bann area. Linear styles represent careful exercises in geometrical drawing, in which the compass played a large role. But if linear styles are somewhat scarce on metal, they are nevertheless plentiful on the trial pieces found at Loughcrew, where, in a rather dismal and dark workshop in cairn H, craftsmen turned out numbers of patterns which are unknown on metal. Without these bone trial pieces and the Turoe stone, Irish art would lose much of its impact. The doodles on the Loughcrew trial pieces are in the linear style, though there is the possibility that these could be translated into the repoussé style should the need arise.

Most of the Irish objects have no associations, and some, like the Petrie Crown, have no localities either. So that, quite often in Ireland, one meets the double disadvantage of unlocalisation and nonassociation. Even the Turoe stone was moved, but luckily not very far. In this chapter we shall use the Turoe decoration as a starting point for further discussion; for many of the objects noted below follow it in time, whilst their decoration owes much to the motifs and patterns represented at Turoe. Even the technique of representation on this stone pillar somehow survives in the repoussé style, and may have suggested it. Because of the lack of associations, dating is a perilous undertaking, and can only be arrived at by analogy. But we have the Galloway-style evidence to draw upon when making our decisions. Whether or not the appearance of the repoussé style across the channel is a measure of contemporaneity cannot be decided upon from the evidence at hand: but in that era preceding Agricola's stationing of his men along the western coast of Galloway, there are grounds for believing that there was a free exchange of ideas between both sides of the North Channel.

If repoussé is the technique influenced (as we think) by the style of decoration on the Turoe stone, then remarkably little of it remains in Ireland. A good example is the Lough-na-Shade trumpet mouth decoration (Fig. 13:3), but the style here is clearly later than that represented on the Torrs pony-cap. A doorknob-type spear-ferrule bears a capable representation of the cable motif;[9] otherwise the only objects of significance left to us, bearing repoussé-style decoration, are the so-called sacrificial dishes, of which two are illustrated here (Fig. 5:2 and Fig. 33:2). Of the two, only the first is representative of a well established definitive art style; it is the finest example of its class, though without provenance. The decoration includes well executed trumpet forms in the elongated Irish style; and it will be seen that here the style is becoming tenuous, as it was on the Stichill collar. We are getting near to the period of devolution. There are several of these recessed bronze discs, but none is so good as this one, for most, like Fig. 33:2, have a degenerate form of decoration, even though it is still in the repoussé style. This makes it clear that repoussé work as such was on the way out, and no real attempt was being made to delay its departure. The translation of patterns into another technique could have come about roughly in the middle of the first century AD, when the boss style elsewhere was coming into its own. An object combining the repoussé and the boss and the boss-and-petal styles is that found at Annalore, Co. Monaghan[10] (Fig. 33:1), which was in association with an enamelled bridle-bit. There are two patterns represented here: of the first there are four bosses linked by S-scrolls in repoussé, having mid expansions in the form of raised quadrants. Quadrants such as these are peculiar to metalwork wherever Irish influences were being felt; and we see one such evolving on the Monasterevan disc (Fig. 33:2). They are found also amongst the decoration on some of the massive armlets, and again on the boar's head from Deskford. Return influences were felt in Ireland, as may be seen by reference to the second pattern which is represented on the Annalore disc. This is the centre-piece, in boss-and-petal style, for which the nearest parallel is the stark decoration on the Stanhope, Peeblesshire, mount (Fig. 56:8). The Stanhope mount was found in association with a massive armlet and a Roman patera, both of second-century date. And though Annalore may be some miles from Newry, the datable link between the Scottish and the Irish areas must surely be the Newry massive armlet (Fig. 9:1).

But, apart from these similarities, it must be clear that Irish art differs from the British, yet both must be clearly labelled Celtic. In Britain there are no equivalents to the Petrie Crown, or to the Cork Horns: neither can the patterns on the Loughcrew trial pieces[11] (Fig. 34) be matched in British art. It is not easy to date these drawings, though immediately obvious is the preoccupation with broken-back scrolls. It is difficult to decide whether these patterns are really Irish, or whether they are continental. This may seem

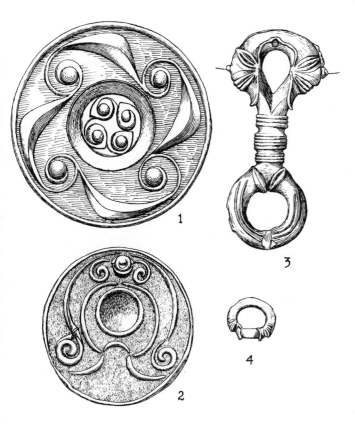

Fig. 33: 1. Decorated Bronze Disc, Annalore, Co. Monaghan. 2. Decorated Bronze Well-disc, Monasterevan, Co. Kildare. 3. Bronze Mirror Handle, Ballymoney, Co. Antrim. 4. Bronze Swivel-ring, A'Chrois, Tiree and Vallay, North Uist. (1/2) except 2 (1/6)

Fig. 34: Loughcrew, Co. Meath: Patterns on Bone Trial Pieces.

surprising; but it is also equally difficult to decide if they were meant for linear decoration or for repoussé work. If for the former, then no metalwork bearing one or other remains. The patterns are otherwise wholly unknown in these islands. Cable motifs figure prominantly in Fig. 34:2 and 3, and both can be equated with the decorative style of the Llyn Cerrig Bach plaque. The yin-yangs of Fig. 34:1 can be paralleled on the central boss of the Battersea shield.[12] Fig. 34:5 comes nearest to the broken-back scrolls of the Lochar Moss beaded torc. The rosettes of Fig. 34:9 are similar to others at Stanwick;[13] whilst the quatrefoil pattern of Fig. 34:12 is not too uncommon in Britain, occurring as it does on the Holderness bridle-bit (Fig. 26:1), and on the terret from Seven Sisters, Upper Dulais, Glamorgan (Fig. 25:3). It also occurs on the petal-shaped strap junction from Middlebie, Dumfriesshire;[14] and lastly on several later Brigantian dragonesque brooches. The centre stud, with criss-cross pattern, resembles others on the helmet from north Britain.[15] All these similarities and parallels point to close connections between the peoples on both sides of the Irish Sea; and, moreover, they are sufficient to suggest a first-century date for the Loughcrew trial pieces. It is not clear whose was the greater influence; but on past records it would appear that the balance is about equal.

The patterns on the Loughcrew trial pieces give no indication of an impending change of style, which in time happened as a result of the dropping of the repoussé technique for another, that of casting objects and patterns together in a single operation. The best known examples of the new style are the Bann disc, the Petrie Crown and the Cork Horns. The first and the last named were found at opposite ends of the island. On looking round for introductory styles and patterns, one must assume that these are present on the disc of Fig. 5:2, with its carefully and beautifully represented trumpets, which are yet not so elongated and so spidery as those on the Bann disc, the Petrie Crown and the Cork Horns. But the metalworkers were careful with their new style, there were few liberties taken with the overall pattern, and, considering the articles in question, it could have been the prerogative of one workshop. Horns were made by casting the metal flat, rolling it into its characteristic pointed tubular shape afterwards. Had the decoration been carried out in repoussé style, this operation would have been impossible. As to the decoration, in all cases the trumpets have become emasculated, so much so, in fact, that they appear to be little more than spidery scrolls. It is clear that inspiration for the pattern on the Cork Horns[16] (Fig. 13:5) must have come from that on the Battersea shield. There is a distinct connection between the arrangement of the horn's decoration and the palmette form utilised on the shield. The Irish trumpet form came into existence because in Ireland all scroll terminations, which also includes all spiral terminations, were becoming swollen. And the eventual trumpet form was probably suggested because trumpets happened to be in vogue in England in the first century AD.

Perhaps the Petrie Crown is later than the Cork horns. This is a supposition that could be turned into fact by the occurrence on the Petrie Crown of hippocampic heads, which in every way resemble those on Brigantian dragonesque brooches of the early second century; and which, at Milking Gap, Northumberland,[17] can be dated not later than 122. These heads (Fig. 35:1) are such true copies of the Brigantian specimens that they appear to have been done locally. This cannot be the case; but somebody in Ireland may have acquired a Brigantian dragonesque brooch, and so become acquainted with the style. Such influences could have crossed the North Channel between 105, following the Roman withdrawal from the Lowlands,[18] and the building of Hadrian's Wall in 122. The rest of the decoration on the Petrie Crown is in much the same style as that on the Cork Horns; but there has been a change of pattern in the case of the Bann disc (Fig. 35:2). A switch from a palmette-based design to one based on the triskele was made. On the Bann disc, decoration consists of little more than a whole series of scrolls with finials in the form of hippocampic heads, still of the same form as those on the Petrie Crown, but now breaking up into isolated units. Perhaps this indicates an over-familiarity with the Brigantian style of head, thereby producing a natural reaction towards abstraction. Possibly a date around the middle of the second century would be appropriate for this object, which is one of the nicest pieces to have come from northern Ireland.

This suggested date would be too late for trumpets in England, and, in any case, their appearance had long since been forgotten at this period of time. Thus we come face to face with the sort of persistence that is frequently noted in Irish art, and which makes dating difficult. The kind of situation with which one is confronted concerns the Navan brooch (Fig. 35:3) in Ireland, and the Mortonhall scabbard in Scotland. Identical trumpets are common to both, and the arrangement is the same. The scabbard (Fig. 32:5) is Brigantian, made possibly before AD 74, but decorated in the north somewhat later, with decoration in the Galloway style. The Galloway style had already ended before 122, when Roman punitive measures were undertaken in that area. So, now the position is this: would an equal date be appropriate for the Navan brooch? Is is impossible to answer either in the positive or the negative.

Fig. 35: 1. Development of Decoration on Horn, Petrie Crown. 2. The Bann Disc, Coleraine, Co. Derry. 3. 'Navan'-type Brooch, Navan Fort, Co. Armagh. 1 (1/2); 2 (1/2) and (3/2); 3 (3/4)

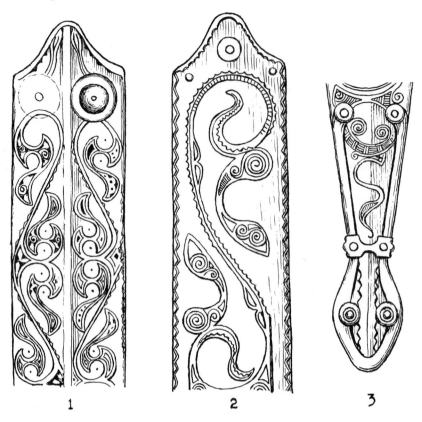

1   2   3

Fig. 36: Decorated Bronze Sword Scabbards, Lisnacrogher, Co. Antrim. (3/4)

The most interesting, if not the best linear decoration occurs on the Lisnacrogher and Bann scabbards. By any reckoning, these are remarkable productions. Piggott[19] includes all Irish swords and scabbards in his Group IIIa, which he derives from the Bugthorpe type (Group III). But this must be a mistake. The Bugthorpe chape is bifid, whereas there is more than a passing resemblance between the Lisnacrogher chapes (Fig. 36:3) and those of the Hunsbury type (Group II). Compare Fig. 36:3 with Fig. 32:1. In the previous chapter it was noted that Group II swords and scabbards were confined to that part of the country south of the Jurassic Ridge, with small concentrations on the south side of the Bristol Channel and in Norfolk, but that the greatest number had been found in the Thames Valley. Contact with the south-western school seems assured. On the other hand, the Bugthorpe form is eastern rather than western, with a decoration relationship existing with the Stanwick patterns and the broken-back scrolls of the Santon mount (compare with Fig. 24:1 and 5). It has already been shown that one pattern at Lisnacrogher (Fig. 36:1) can be paralleled at Stamford Hill; whilst similarities in the St Mawgan and Lisnacrogher lotus-bud motifs (Fig. 10:1 and 2) have also been commented on. A strong link with the mirror decoration, also of southern form, is established in the basketry filling that also occurs at Lisnacrogher. So that, when all these bits of information are gathered, it will be seen that at Lisnacrogher they add up to a leaning on south-western art forms, which, with the form of chape, more than suggests a link with the Group II scabbards of England, and the art of the south-western school. Therefore, Fox's[20] idea of the existence of a close relationship between Bugthorpe and north Irish art is no longer tenable. A third scabbard from Lisnacrogher (Fig. 36:2), seldom published, has decoration which can be paralleled on the Torrs drinking-horn mounts.[21] The only recognisable motif is the cable, here with spirals at the circular end, heralding, or suggesting the snail-shell roundels of the Broighter torc: or, conversely, one is a degeneration of the other. Spirals of this form possibly were suggested by the in-turning of the tendrils on the lotus-bud motifs, as seen on another Irish scabbard, this time from the River Bann at Toome[22] (Fig. 13:6).

One may conclude, with some justification, that there were two artistic traditions in Ireland. This situation can be explained by the fact that, though initially both had the same ancestry, nevertheless they are representative of two separate introductions into the country in two separate periods. On the one hand, there is the art as represented on the Turoe stone, with a tradition in Ireland perhaps going back to that Waldalgesheim phase represented only by the Clonmacnoise gold torc, and which, for the most part, belongs to the central belt of Ireland. On the other hand, there are the contrasting linear styles of Lisnacrogher, introduced at a much later period, perhaps not before the beginning of the first century AD, and coming up channel from St Mawgan and the south-west at a time when linear styles only were known in those parts.

Apart from the theorising, with the occasional fact added, there is little more that can be said; except to add that the second century would appear to have witnessed the going down and out of all the artistic skills noted in this chapter, leaving little behind but the skeleton of an art movement, a hotchpotch of badly formed debased motifs or their remnants, spirals alone being left to further progress. Yet manufacturing skills were in no way affected. The further progress of what remained during the centuries that followed will be dealt with in succeeding chapters.

NOTES

1. W.F. Wakeman, 'On the Crannog and Antiquities of Lisnacrogher, near Broughshane, Co. Antrim', *Journ. Roy. Soc. Antiq. Ireland*, IX (1889), p. 97.

2. David Greene, 'The Chariot as described in Irish Literature' in C. Thomas (ed.), 'The Iron Age in the Irish Sea Province', *CBA Research Report*, no. 9, p. 61.

3. Ibid., p. 71.

4. R. Haworth, 'The Horse Harness of the Irish Early Iron Age', *Ulster Journ. Arch.*³, XXXIV (1971), p. 26.

5. Jope, 'The decorative cast Bronze Disc from the River Bann, near Coleraine', p. 95.

6. Armstrong, 'The la Tène Period in Ireland', p. 14, Fig. 9.

7. Eoin MacNeill, 'The Pretanic Background in Britain and Ireland', *Journ. Roy. Soc. Antiq. Ireland*, LXIII (1933), pp. 1–28.

8. Duignan, 'The Turoe Stone', p. 206.

9. J. Raftery, *Prehistoric Ireland*, Fig. 253:13.

10. Ibid., Fig. 266.

11. H.S. Crawford, 'The Engraved Bone Objects found at Lough-Crew, Co. Meath, in 1865', *Journ. Roy. Soc. Antiq. Ireland*, LV (1925), pp. 15–29.

12. Fox, *Pattern and Purpose*, pl. 16.

13. *BM Guide to Early Iron Age Antiquities*, p. 142, Fig. 161.

14. Wilson, *Prehistoric Annals of Scotland* (1863), p. 458 and illustrations.

15. Fox¹², pl. 62:C.

16. M.J. O'Kelly, 'The Cork Horns, the Petrie Crown and the Bann Disc', *Journ. Cork Hist. and Arch. Soc.*, LXVI (1961), p. 1.

17. Kilbride-Jones, 'The Excavation of a Native Settlement at Milking Gap', p. 342, Fig. 5.

18. Sheppard Frere, *Britannia*, p. 144.

19. Piggott, 'Swords and Scabbards of the British Early Iron Age', p. 14.

20. Fox¹², p. 43.

21. Ibid., pl. 20.

22. Jope, 'An Iron Age Decorated Sword Scabbard from the River Bann at Toome', p. 81.

14. The Soghain and the Cruithin

In Britain, during the period with which we are dealing, the names of tribes and their territories are known. Recurrent names are those of the Brigantes and the Votadini, not to mention the Caledonians, who were not a tribe in themselves, but a number of tribes living (mostly) north of the Tay. We are familiar with the name of Boudicca, Queen of the Iceni, and of the rebellion that cost the lives of some 70,000 people. We know about the marital squabbles of Cartimandua and her consort, and about the varying fortunes of tribes whose misfortune was to take up arms against the Romans. Tacitus makes special mention of the Silures and their sallow skins, supposing them to be Iberian. In Ireland, historical contact is less common, and also it is less historical.

In the course of this study, two Irish tribes have been mentioned, because, in one way or another, objects of art were found, or had their being, in their territories. These tribes are the Soghain and the Cruithin, the latter being related to the former. According to Mac Fir Bhisigh there were seven sub-divisions of the Soghain. Unfortunately, not very much more is known about these people, except that they lived in parts of the present counties of Galway and Roscommon, on both sides of the River Suck.[1] Others of the same tribe lived in Meath and Monaghan, and there were colonies at Bantry and in the Muskerry country, to the west of Cork city. The Soghain were in possession of these midland territories until they were displaced by the Sept Ui Maini in the fifth century AD.

In existing references to the Soghain they are described as the ruling kindred of the regions named. This is a most interesting statement, because it suggests that the Soghain were an aristocratic minority of princes and nobles. O'Donovan gives an account of the displacement of the Soghain in the *Book of Rights* (p. 106) in which he makes reference to their king as Cian 'the firbolg king of the district'. Genealogical Tracts I, 75 mentions the 'ancient kin of the old plain of the Soghain', from which it could be inferred that the Soghain were of ancient lineage. But, how ancient? Were they in fact settlers from the Continent, as suggested by the presence of the Clonmacnoise gold torc?

O'Donovan's reference to Cian as 'the Firbolg king of the District' is interesting in another context. The ancestor of the Firbolg is given as Nemeth, and, according to the Book of Invasions, the people of Nemeth were engaged in trade with the Greeks, by exporting 'Irish Earth' in bags — hence the name Firbolg, bag men. It seems that the Greeks used to spread this earth around the perimeters of their cities, as a protection against venomous reptiles!

Nemeth's brother was by name Partholon. Now Partholon was the ancestor of the Cruithin, that is, Picts. The same account states that there was 'an original population of Ireland consisting of Cruithin and others who are comprehensively called Fir Bolg.' But if the bag men were in fact the Soghain (as it would appear), then it is they, and no one

132

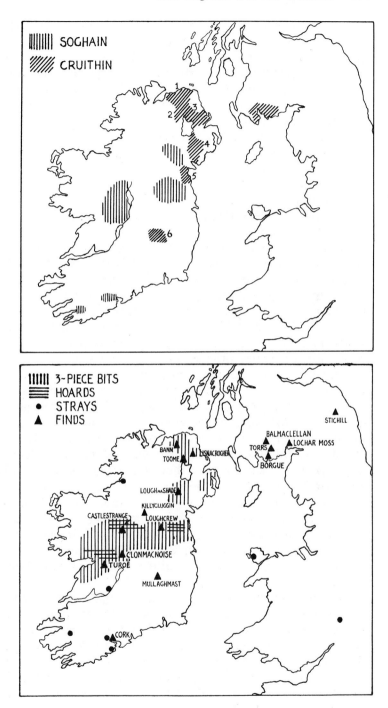

Fig. 37: *Top*: Map showing Lands occupied by the Soghain and the Cruithin.
Bottom: Distribution of Irish Three-link Bridle-bits and Locations of Celtic Art Objects.

else, who were in contact with the Greeks. Such contact would have brought them in close touch with continental cultures, and this could be the information needed by way of explanation for the presence of the best works of art in 'the old plain of the Soghain'.

On the other hand, the Cruithin appear to have been contented with staying at home. Most of their autonomous territories were in eastern Ulster, as is indicated on the map, Fig. 37. There were seven, maybe nine petty kingdoms of the Cruithin, and these still existed as late as AD 563. The known territories are listed below, the numbers corresponding with the numbers on the map:

1. Arda Eolairg
2. Léi (taken from the Cruithin by the Ui Neill in 563)
3. Northern Dál Araidi
4. Southern Dál Araidi
5. Conaille of Muirtheimne
6. Lóigse (Leix).

MacNeill states that there were Cruithin settlements in the Lough Mask area, and that there were subject populations of Pictish origin in Co. Down and Co. Longford. There were also Cruithin in Galloway, in the territory of the Novantae. Here they retained their identity long enough to earn a place in written history, for some were converted to Christianity by St Ninian, about the beginning of the fifth century. St Patrick referred to them as 'the apostate Picts'; but subsequently they disappear from history. It is perhaps worth mentioning that when Agricola invaded the territory of the Novantae in AD 82 he found there tribes hitherto unknown by name.

The Soghain occupied a corridor of land which stretched roughly from Dundalk Bay to Galway Bay. The Cruithin occupied another strip of land from south of Lough Neagh to the mouth of the Bann. As 'the ancient kindred', the Soghain could have been descendants of Bronze Age aristocratic families; or they could have been Iron Age aristocratic incomers, as put forward above, and still be regarded as 'ancient kindred' by the later chroniclers. It is also clear that they were horsemen, judged by the large number of bridle-bits[2] that have been found in their territories. Hoards of three-piece bits are common in counties Galway, Roscommon and Meath. No hoards have been found in the territories occupied by the Cruithin, only single finds. Stray bridle-bits are rare: there is one in Co. Donegal and another in Tipperary, whilst a few congregate round Cork city. These finds near Cork cause no surprise, seeing that a Soghain colony once existed here. But the main concentration remains in Galway, Roscommon and Meath. The reduced numbers of bridle-bits found in Cruithin territories suggests that they may have got their bits from the Soghain.

This evidence of equitation is interesting, since riding on horseback was completely unknown to Irish civilisation, a fact indicated by there being no native word for 'trousers'.[3] As David Greene has remarked: 'in this respect the Irish are in remarkable contrast to the Celts of *Gallia bracata*'. So what can be made of this conflict of evidence? It would seem that the second possibility put forward above is the correct one, and that the Soghain were aristocratic incomers who superimposed their authority upon the Irish, who at that time may have still maintained a late Bronze Age form of economy.

Howarth regards the Irish bridle-bit as an imported form, the whole series stemming from a limited group of imports at one time from one place.[4] His type A can be matched at Llyn Cerrig Bach,[5] where it was found in a hoard which some regard as a votive offering, but others as a deposit by first-century refugees fleeing before the Romans, as did those other refugees who settled on Lambay Island. Information is scarce, and not easy to come by, but what there is tends to suggest that Irish bridle-bits were made about the turn of the century BC–AD.

Jope also has proposed a derivation of the Irish bridle-bit direct from the Continent, suggesting the Marne region.[6] An earlier import into Soghain territory is the buffer-ended gold torc from Clonmacnoise,[7] the only undisputed example of la Tène work in the Waldalgesheim style to have been found in Ireland and Britain, and generally regarded as being of third-century-BC date. Partholon, Nemeth's brother, is known to have landed near the mouth of the River Erne, and from Ballyshannon comes an anthropoid hilt of la Tène III date.[8] The suggestion that these objects were imported direct from the Continent is boosted by the occurrence on the Turoe stone of a whole repository of continental art motifs and patterns. Local furtherance of this art style is recorded on the stones from Castlestrange,[9] Killycluggin[10] and Mullaghmast.[11] Turoe and Castlestrange are situated on the old plain of the Soghain. Mullaghmast is in Lóigse. And although Killycluggin may appear to be outside the Monaghan area, Soghain may have lived here too. The dates suggested for these stones are various: in fact, anything from the third to the first century BC. But the maximum period of activity could have been from about the middle of the first century BC to the middle of the first century AD, during the second half of which metalworkers set up shop in cairn H at Loughcrew, in the Soghain lands of Meath.

Origins are never easy to trace; but everyone should now agree that there is sufficient evidence to connect the Soghain with early artistic activity in Ireland. The techniques and skills displayed in the making and decorating of metalwork are the equal of others anywhere, and sometimes superior. Again these skills look as though they were implanted, rather than that they grew out of the long tradition of Bronze Age metalworking in Ireland. Art forms were in a high state of development when first they appear, and such skills are not learned overnight. Workers with these skills are most likely to have been settlers.

The Soghain type is a repoussé style, and seemingly suggested by the rounded off forms of the Turoe stone patterns. The ingenuity in devising patterns as noted here still existed at Loughcrew; and the style almost penetrated Cruithin territory (at Lough-na-Shade) and continued by being the foundation of the Galloway style. Balmaclellan, Lochar Morr and Stichill are examples of metalwork displaying later developments, and all demonstrate a continuance of artistic contact with Irish sources . . . the similarities in the patterns on the Cork Horns and the Stichill collar should be sufficient proof. Only the Cruithin of Co. Antrim appear to have looked elsewhere for inspiration; or was it that amongst them there were late comers from Armorica?

In this short account it was not our purpose to prove anything, only to put before the reader points for further thought. At any time it is worth retrieving from the mists of antiquity the names, however shadowy, of tribes within whose territories all the work was done, and to highlight the achievements attained in metalworking and pattern-making. The story about selling Irish earth to the Greeks could be wholly fanciful, but it is still worth mentioning, since there was an earlier Greek connection. From Cloonlara, Co. Mayo;[12] Annadale, Co. Leitrim;[13] Knock, Co. Mayo;[14] and Clonbrin, Co. Longford[15] come Herzsprung shields[16] of Greek origin, but of Bronze Age date. Two arrived via Spain, whilst the other two arrived via Scandinavia. This evidence may add little or nothing to what we have been talking about, but it does serve to prove early contact with the Mediterranean area. Also these shields tell us that the point of ingress has always been the west of Ireland, as it was in Neolithic times when Killala Bay was chosen for the introduction of court cairns. This seems to be contrary to geographical reasoning, yet the same area, that is the mid west of Ireland, again witnessed the beginnings of what is known as Celtic art. The role played by the Shannon is not clear. But, so far as the Soghain are concerned, what is clear is that the Soghain colonies at Bantry and at Cork were, in fact, trading stations.

NOTES

1. Eóin MacNeill, 'The Pretanic Background in Britain and Ireland', p. 1.

2. Haworth, 'The Horse-Harness of the Irish Early Iron Age', p. 26.

3. Greene, 'The Chariot as described in Irish Literature' in Thomas (ed.), p. 61.

4. Haworth[2], p. 31.

5. Fox, *A Find of the early Iron Age from Llyn Cerrig Bach, Anglesey*, no. 55.

6. Jope, 'An Iron Decorated Sword Scabbard from the River Bann at Toome', p. 88.

7. Armstrong, 'The la Tène Period in Ireland', p. 14, Fig. 9.

8. Megaw, *Art of the European Iron Age*.

9. Armstrong[7], pl. VI.

10. R.A.S. Macalister, 'On a Stone with la Tène Decoration recently discovered in Co. Antrim', *Journ. Roy. Soc. Antiq. Ireland*, LII (1922), p. 113.

11. Armstrong[7], pl. VI.

12. A. Mahr, 'New Aspects and Problems of Irish Prehistory', *Proc. Preh. Soc.*, 3 (1937), p. 262, pl. 25.

13. G. Coffey, *The Bronze Age in Ireland* (Dublin, 1913), p. 75, Fig. 69.

14. S.P. O'Riordain, 'Prehistory in Ireland', *Proc. Preh. Soc.*, 12 (1946), p. 142, pl. 14:2.

15. Coffey[13], p. 77, Fig. 70.

16. H. O'N. Hencken, 'Herzsprung Shields and Greek Trade', *American Journ. Arch.*, LIV (1950), p. 303.

PART THREE: POST-INVASION STYLES

15. Introduction

Brigantia, as was noted in Part II, was doing well as a buffer state between the Romanised south and the barbarian-occupied north; and even a new town had sprung up. Expansion was afoot, and seemingly even the craftsmen were doing well. Under the circumstances everyone might have hoped that Venutius would stay his hand and behave himself. Since his quarrel with his former wife (Cartimandua had divorced him some time after AD 60) he had managed to keep the Romans at a distance, and this was still the situation in AD 71. Yet, by AD 74 Brigantia was battered and leaderless. Shortly after, Wales was reduced to a vassal state. In 78 Agricola arrived in Brigantia, and set about policing the country; and then, in 79, he began his advance northwards, and he reached the Tay a year later.[1]

Current events must have seriously affected the livelihood of the metalworkers. The putting down of the Icenian rebellion led to the burying of trappings. With the increasing Romanisation of the country, native styles and native metalwork was being rapidly phased out, particularly in the Civil Province. This is reflected in the art. There was a general simplification of patterns, with abstraction carried almost to extremes. The mirror style broke up. Repoussé styles were wilting. Large objects of almost any nature were fast disappearing. This was the general situation in the period covered by Part Two of this present book. In Part Three the process continues. Only in the north were bridle-bits and terrets still being made, and Caledonia had an industry of its own for a while. Votadinian friendship for Rome was about to pay off, with Agricola's bypassing Traprain Law, on his way to suppress the Caledonians; and submission brought the builders of the Orkney brochs into contact with the friends of Rome.[2]

Robbed of their markets for the bigger things, the metalworkers now turned to the production of smaller articles, mostly of a personal nature. The roads which the Romans had laid down aided travel, and so made for easier trading and quicker deliveries. Traprain got busy with dress-fasteners, and further south everyone seemed to be making brooches. But trouble was never far distant from anyone at this time; and possibly as a result of Agricola's harsh treatment of hostile tribes, some forts went up in flames at the start of the second century, forcing the Romans to evacuate southern Scotland in headlong haste. The frontier was withdrawn to south of the Cheviot, giving the northern tribes a whole generation's breathing space. The Romans did not return until 122, when the Wall between the Tyne and the Solway was commenced, as a frontier against the warring tribes to the north; and it took six years to build. This is known as Hadrian's Wall. After 142 the whole frontier system was moved northwards to the Forth-Clyde isthmus, thereby bringing the Lowland tribes within the Roman aegis. The new situation is reflected at Traprain Law, for now the very coarse native ware is replaced by Roman, mostly Antonine pottery. Figured Samian was imported from Gaul. Second-century coins make an appearance. The picture here is one of increased prosperity, once more emphasising the privileged position enjoyed by the Votadini, in which they were permitted to carry on with their own way of life, with freedom to cultivate grain and to tend their herds.[3] In Wales, a similar privileged position was enjoyed by the people of Tre'r Ceiri.[4]

Trouble was frequent in the military areas. In Brigantia another revolt broke out in 158. Collingwood[5] blames this revolt for putting an end to the trade boom in brooches. But worse was to follow: early in the reign of Commodus (176–92) Roman-held territory was again invaded from the north, with disastrous results. The Antonine Wall was abandoned, and in 196 the situation further worsened when the governor, Clodius Albinus, transferred the Roman garrisons to the Continent for his attempt (later unsuccessful) to gain the purple. The result was that the northern tribes were able to burst into a virtually defenceless province, plundering and pillaging as they went, and taking their trail of destruction as far south as York. This time, Traprain Law did not escape: the workshops were destroyed, and the same was true of the Brigantian workshops; Philo-Romans and their property suffered most. But these events cannot explain why the factory on Nor'nour, an islet in the Scilly Islands, was destroyed at the same time.

For our metalworkers, the position could hardly have been worse. Now their workshops lay in ruins, and their trade was gone. The art went underground. Presumably, all had to pay the penalty for trading with the Romans. The disaster is interesting for one good reason: the cessation of production proves finally that all the workshops were in the fringe areas. Of all the forms of brooch produced, there was but one lone survivor, the zoomorphic penannular brooch. It survived because the form had been taken to Ireland. Ireland, naturally, remained unaffected by these events, and so did the Caledonians, but by this time their metalwork had ceased to be important. Alone, and outside this sea of disaster, the Irish metalworker was left to carry forward a waning tradition.

The Romans did not return until 208. In 209 the Caledonians were reduced to unconditional surrender. Once more Roman coins circulated freely in the Lowlands.[6] But, never again were conditions the same. Trade was severely reduced, and there was now a greater dependence on Roman-made products. Tribal fortifications were prohibited, except in the case of Traprain Law, which, after lying unoccupied for some 30 or 40 years, appears to have assumed an increased importance. Even the workshops got going again, but now manufacture was on a greatly reduced scale. Products were mostly pins.

Under the Severan reorganisation, recruiting for frontier duties took place from amongst local tribesmen. Natives set up dwellings close to the Wall, and many of these must have housed the wives and children of these British auxiliaries.[7] As a result, the duties of frontier defence were transferred to the Votadini and the Dumnonii. This Romanising influence doubtless affected the manner of dress of the times. Native dress may have been proscribed, so that dress-fasteners and native-made brooches became things of the past.

Now it was the turn of the Irish to create trouble. Irish raids on the south-west coast of Wales were stepped up round about AD 275. An Irish dynasty and an Irish aristocracy settled in Pembroke, and men from Leinster occupied western Caernarvon. To counter this threat, a new fort was built at Segontium, replacing the old; and another fort was built on Holyhead Island. Their purpose was the protection of the valuable Anglesey copper mines.[8] There were also Irish raids on Morecambe Bay, so a new fort was built at Lancaster. The year 296 brought trouble along Hadrian's Wall, and the garrison was withdrawn. Immediately, the tribes invaded and plundered. But, this time swift retribution followed, and all Roman outposts had been rebuilt by 305. From this period, all trade between the areas on both sides of the Wall had to be conducted through the Customs Post at Knag's Burn.[9] Discrimination against northern products is now evident. Although southern-made iron tools, coarse pottery and glass were in use at Traprain Law, nothing was exported from the north in return.

There was a general alarm in 343, which could have been accounted for by another Irish raid. Then, in June 367[10] came the *conspiratio barbarica*, in which the Irish in the west, the Picts in the north, and the Saxons in the east combined together in concerted raids over a large area of the country. All forts and civilian settlements outside the Wall area were destroyed. Corbridge and Malton suffered severe damage. Rebuilding was hurriedly carried out, yet out of order came disorder. In Wales Segontium was rebuilt once more, again to protect the copper mines. In Welsh legend, Segontium is associated with Magnus Maximus;[11] but Maximus perished in 388, and from this time until the rescript of Honorius, the Irish began to pour into north Wales and to settle there.

These are the events that led up to the Withdrawal. After the Romans had gone, Celtic art got a fresh start, but it had been a gradual process, with its beginnings noted during the Celtic revival at the latter end of the fourth century. The rest will be a matter for Part Four of this book. In the course of Part Three most of the items discussed are pre-196: from that year onwards there is something of an artistic vacuum. It is curious to relate that it was the British, and not the Romans, who brought about this situation.

NOTES

1. Sheppard Frere, *Britannia*, p. 125.
2. V.G. Childe, *Scotland Before the Scots* (Methuen, 1946), p. 129.
3. J. Curle, 'An Inventory of Objects of Roman and Provincial Roman Origin found in Scotland', *Proc. Soc. Antiq. Scot.*, LXVI (1931–2), p. 344.
4. *Arch. Camb.*[6], IV (1904), p. 9, Fig. 6.
5. R.G. Collingwood, 'Romano-Celtic Art in Northumbria', *Arch.*, 80 (1930), p. 57.
6. A.S. Robertson in *Proc. Soc. Antiq. Scot.*, LXXXIV (1949–50), p. 156.
7. K.A. Steer in I.A. Richmond (ed.), *Roman and Native in North Britain*, p. 108.
8. E. Birley in *Arch. Ael.*[4], XIV (1937), p. 172–7.
9. Birley[8], p. 172.
10. R. Tomlin, 'The Date of the Barbarian Conspiracy', *Britannia*, V (1974), p. 306.
11. Richmond[7], p. 121.

16. Beaded Torcs

The beaded torc is a positive survivor from an earlier heroic age, and one would need to be a stoic in order to wear one. The form differs from all other torc forms: roughly half of the circle consists of a solid bar, which in cross section can be square, rectangular or round, whilst the remaining half of the circle is made up of heavy metal beads, strung on to a circular metal bar of small cross-sectional area. This bar is detachable. Generally, the whole torc is decorated, though plain specimens are known. Torcs of this form are not very numerous, and their distribution is confined to the northern half of England, with strays in north Wales, Worcestershire and south-west Scotland. They may have been Brigantian in origin, though probably the original idea was Roman. In all cases they are made of bronze.

Frank Simpson[1] has shown that torcs were a widely diffused Roman military decoration. Individual acts of valour were rewarded by the giving of torcs. Inscriptions record awards to legionaries and private soldiers. In the cavalry, the torc was a subordinate officer's badge. One such torc, plainly not of British manufacture, was found at Benwell.[2] It is made up of a curved double-flanged girder-shaped strip and square beads with swollen rounded-off corners, the beads being separated from one another by tubular spacers. The sole decoration is a light chevron pattern, which follows the base of the outer flange. This is plainly the pattern for native torcs.

The sharp edges of the Carlisle torc must have severely tested the endurance of the hero whose reward it was. A torc similar to the Benwell specimen was found at Perdeswell, Worcester.[3] Although it is already becoming clear that there is no evolutionary series, nevertheless torcs divide into two types: the first is one based on the girder-shaped bar, whilst the second has a rounded bar for the solid part of the hoop, which, in the latter case, tends to occupy roughly two-thirds of the circle, as against half for the girder form. The round-bar type usually has round beads, whereas the girder type has either melon-shaped beads, or they are square with the corners rounded off. Some melon beads have swirled ribs: this type seems to be peculiar to the west country, whereas the rounded-bead, round-bar type has an eastern distribution. There may be no particular significance in this distribution, since a round-hooped type was found at Carlisle, whilst Benwell is on the east coast.

The best known beaded torc, and perhaps the best, is the one from Lochar Moss (Fig. 38), found two miles to the north of Cumlangan Castle, Dumfriesshire.[4] Clearly it has been fashioned after a Roman model, by craftsmen still steeped in the traditions of the Galloway style; and by adapting what is essentially a non-Celtic form they gave to it a thoroughly Celtic look. Fox's view that the Lochar Moss torc is in true northern style[5] can be qualified further by localising the style to Galloway. The curved girder-like bar is

Fig. 38: Decorated Bronze Beaded Torc, Lochar Moss, Dumfries. (3/4)
Below: Detail from Decoration on Trial Piece, Loughcrew, Co. Meath.

in the Benwell tradition, whilst the 14 beads (some are missing) have been threaded on to a circular bar of small diameter, on which they are kept apart by collars. These spacers are pulley-shaped, a useful weight-saving idea. The shape of the beads is unusual, in that it is not Celtic but Roman, after the melon-bead form. The beads are cast, which is why one resembles another closely. The girder-shaped bar is decorated at the front with running broken-back scrolls, associated with bosses which, in effect, are the heads of rivets, used here to retain in position these same scrolls, which have been shaped separately in openwork. The fact that decoration has been applied in this manner, and kept in position by the rivets, is a unique labour-saving idea, noted before in the case of the Llandyssul collar, though here the decoration was cast. Fox[6] describes the Lochar Moss boss-and-scroll work as:

> a familiar form in which each boss serves two scrolls, one arriving, the other departing. The sweep of each scroll here, however, is abruptly closed by a change of direction — Leeds' broken back[7] — through nearly a right angle; and each boss serves, as it were, two half scrolls.

An explanation is called for to explain the presence here of the broken-back type of scroll, which is fully fledged, showing no signs of tentative beginnings. Therefore, the broken-back scroll must have been imported from somewhere, seeing that its beginnings are nowhere to be found in Britain. There are plenty of broken-back scrolls in Ireland, and the artists of Loughcrew were preoccupied with them; and, for our benefit, they drew some well thought-out examples on the bone trial pieces. But, at Loughcrew, they also drew rosettes, of a form seen at Stanwick and Balmaclellan; and by so doing they have told us of the close connections that existed between the art schools of Ireland and those of Britain. An exchange route through Galloway is indicated, even if it cannot be proved. But already we know that it was Galloway 'to which (Irish) merchants resorted for the sake of commerce, the harbours and approaches to the coast being well known'.[8] With Cruithin on both sides of the North Channel there must have been a good deal of travelling to and fro; and why not? At a guess, the Lochar Moss torc must have been made before 82, when Agricola 'lined the coast with a body of troops'. The Galloway style was waning at this time, and it must have ended altogether when the Romans undertook punitive measures here in 122.

However, the rest of the collar is of local design, and it must have been made in local workshops. The bead-ribs are seed shaped, divided by rib-edged grooves. Beads precisely similar were found in the Hyndford Crannog, Lanarkshire,[9] but here the rest of the torc is missing. The Hyndford beads (Fig. 39:2) were associated with pottery and other objects dating just before or just after AD 100.

None of the remaining torcs is so characteristically Celtic. But a clear relationship is apparent between the Lochar Moss torc and the specimen from Mow Road, near Rochdale, Lancashire.[10] In appearance, this torc (Fig. 40) is singularly barbaric, consisting of a heavy rectangular sectioned bar with raised edges, and decorated near those edges with small triangles on two levels, one row of triangles being stepped back from the other. But the reverse side is decorated with running peltae. The curiously shaped beads must have been modelled on the Lochar Moss form, except that here the ribs are swirled. Small pulley-like spacers have been employed to keep them apart.

Three beads (Fig. 42:2) in a more simplified style were found in the native oppidum of Tre'r Ceiri,[11] in north Wales. These beads, which are in poor condition, appear to have been cast together; for, although they are spaced, there appears to be an absence of spacers. Very similar beads, but this time kept apart by means of flat washers, are characteristic of the beaded torc (Fig. 41:3) from Lambay Island, Co. Dublin.[12] This is the first of the round hoops to come up for consideration: and it will be seen that roughly two-thirds of the circuit consists of nothing more than a plain hoop, the beads occupying the remaining third. This simplicity is one of the characteristics of this form of beaded torc, making it distinctly different from the square bar or girder type, on which the division into bar and beads is about equal. The circular form was less of a burden to wear, since it weighed less. Decoration is at a minimum. This Lambay torc came from a small Romano-British cemetery, which yielded other items including dolphin brooches current in Neronian times and until the middle of the second century: also there was a thistle brooch of a type not uncommon in Britain on Claudian sites. On this evidence, the find it most likely to be of mid-first century date,[13] or perhaps a little later. In Hawke's opinion, these people who lived on Lambay were British refugees fleeing before the Roman advance; or alternatively, that the finds were Belgic south-east or Roman loot. This opinion should not be lightly dismissed, since at about this time Agricola was consolidating his grip on Wales, and the local population, particularly the Silures, may have fled, to seek shelter in fringe areas. On the other hand, seeing that the Lambay objects were not buried together as a hoard, but appear to have accompanied individual interments, an established settlement could be indicated. But those who settled here are unlikely to have been Brigantian, as Rynne has made out,[14] for the reason that the decoration on the bronze scabbard mount can be paralleled at Polden Hill, where it can be seen on one of the mounts (Fig. 24:3); and further, British parallels (quoted by Rynne) for the decoration on the gold band are to be found at Stroud, Gloucestershire, and Santon, Norfolk. The spherical-triangular shaped plaque has a parallel at Moel Hiraddug: so that the original premise that these were British refugees fleeing from Wales should still stand. The gentleman who wore the torc was possibly high on the wanted persons' list.

Thus, there are plain hoops, and there are some that are decorated. Amongst the plain there are those from Skerne, East Riding of Yorkshire;[15] Fremington Hagg, North Riding of Yorkshire;[16] and New Mains, Whitekirk, East Lothian.[17] The beads, all cast together, have nothing beyond four incised lines encircling each bead on the Skerne torc; punch-mark filled grooves on the New Mains torc; whilst the Fremlington torc beads are plain. All these British round-hoop torcs have beads that are cast together into a single unit, which is detachable from the hoop. Other torcs, like those from Embsay, Skipton, West Riding of Yorkshire[18] (Fig. 42:1) are more highly decorated, on both hoop and beads. The decoration follows no known pattern in Celtic art, consisting as it does of central wavy lines encircling all the beads, whilst to both sides of this wavy line, and at right angles to it are small petal-shaped hollows. Similar hollows occur, closely spaced, on the hoop, to give a sort of herring-bone pattern. The torc from Carlisle[19] is a little like it, except that cross-hatching has taken the place of these wavy lines on the beads. The hoop is lozenge shaped in section, the two outer facets bearing cross-hatching between two parallel lines. That other Carlisle specimen, from Rickerby Park, Stanwix,[20] is very similar, but the hoop is missing. The specimen (Fig. 39:1) from Lamberton Moor, Berwickshire[21] is badly preserved, with beads cast without the intermission of collars. There are transverse parallel mouldings with cross-hatching both sides on each of the beads. The hoop terminals were once decorated with parallel channelling. Stevenson[22] regards this torc as being typologically the last of its type. It had been passed for scrap when it was buried with a small hoard that contained paterae of Roman origin, probably some time during the Antonine period.

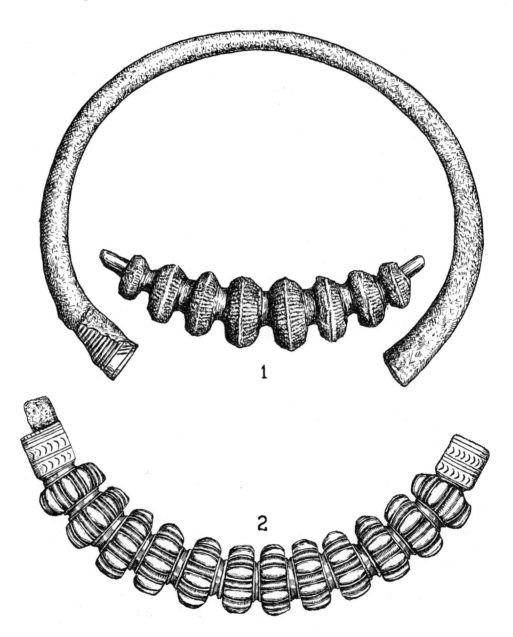

Fig. 39: Bronze Beaded Torcs 1. Lamberton Moor, Berwickshire. 2. Hyndford Crannog, Lanarkshire. (3/4)

Fig. 40: Bronze Beaded Torc, Mow Road, Rochdale, Lancs. (3/4)

There is little positive evidence to draw upon when trying to arrive at a date for these torcs. The Benwell torc was found on the site of a Hadrian's Wall garrison fort; the Stanwix beads were deposited before AD 128; the New Mains, Whitekirk, specimen came from a site which has yielded a fragment of a first/second century patera; but at Lamberton Moor a closer date than the Antonine period is provided by the two head-stud brooches. On balance, a date early in the second century would seem to be appropriate, but against this there is the Lambay Island evidence, which would put this torc back into the second half of the first century AD. Naturally, torcs such as these probably would have been treasured possessions, and as such they would have had a long life. There is just a possibility that the round-hoop type is later than is the girder form, and, from their shape, more comfortable to wear: in which case a stoical attitude would not have been a prerequisite for wearing them.

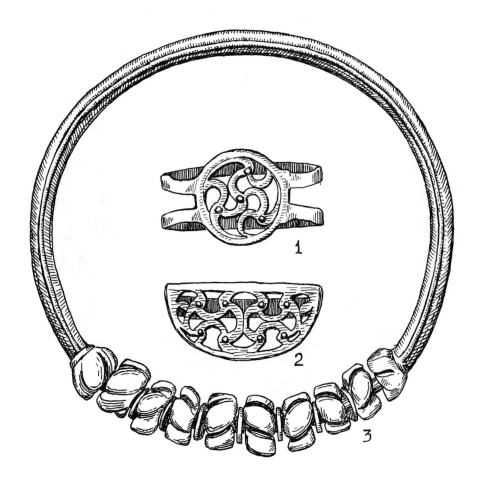

Fig. 41: Lambay Island, Co. Dublin 1. and 2. Scabbard Mounts. 3. Bronze Beaded Torc. (3/4)

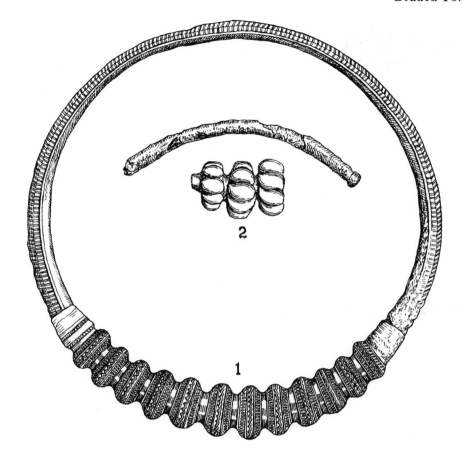

Fig. 42: Bronze Beaded Torcs 1. Embsay, Skipton, Yorkshire. 2. Tre'r Ceiri, Caernarvon. (3/4)

NOTES

1. F.G. Simpson and I.A. Richmond, 'The Roman Fort on Hadrian's Wall at Benwell', *Arch. Ael.*[4], XIX (1941), p. 24.

2. Ibid., pl. ii.

3. J. Allies, 'Ancient Bronze Ornament found at Perdeswell, near Worcester', *Arch.*, XXX (1844), p. 554 and figure.

4. *Arch.*, XXXIV (1852), p. 83, pl. XI.

5. Fox, *Pattern and Purpose*, p. 107.

6. Ibid., p. 107.

7. Leeds, *Celtic Ornament*, p. 52.

8. Tacitus, *Agricola*, XXIV.

9. *Proc. Soc. Antiq. Scot.*, LXVI (1931–2), pp. 381–2. Ibid., LXXXII (1947–8), p. 294.

10. H. Fishwick, 'History of the Parish of Rochdale, 5', *Arch.*, XXV (1834), p. 595.

11. H. Hughes, 'Tre'r Ceiri', *Arch. Camb.*[6], VII (1907), Fig. 2.

12. Macalister, 'On some Antiquities discovered upon Lambay', pp. 240–46. E. Rynne, 'The la Tène and Roman Finds from Lambay, Co. Dublin', *Proc. Roy. Irish Acad.* 76 C (1976), pp. 231–4.

13. C.F.C. Hawkes, 'A Panel of Celtic Ornament', *Antiq. Journ.*, XX (1940), p. 347.

14. Rynne[12], p. 242.

15. Macgregor, *Early Celtic Art in North Britain*, no. 207.

16. Ibid., no. 196.

17. Ibid., no. 206.

18. 'Bronze Collar found at Embsay, in Yorkshire', *Arch.*, XXXI (1846), p. 517, pl. XXIII.

19. Macgregor[15], no. 199.

20. Ibid., no. 208.

21. J. Anderson, 'Notes on a Romano-British Hoard, found on Lamberton Moor, Berwickshire', *Proc. Soc. Antiq. Scot.*, XXXIX (1904–5), p. 367; J. Curle, 'Objects of Roman and Provincial Roman Origin found in Scotland', *Proc. Soc. Antiq. Scot.*, LXVI (1931–2), p. 363.

22. R.B.K. Stevenson in Rivet (ed.), *The Iron Age in North Britain*, p. 26.

17. Craftsmanship in Caledonia

Caledonian was the name given by the Romans to several tribes living north of the Forth-Clyde isthmus, and mostly in the north-east of Scotland. Not only was there a language barrier, but these people differed both racially and politically from all the tribes living to the south, in what is now known as the Scottish Lowlands. 'Caledonia stretches a vast length of way towards the north', wrote Tacitus, adding: 'the ruddy hair and lusty limbs of the Caledonians indicate a German extraction'. A strong central European element amongst these people is possible, in view of the Beaker settlements; but Tacitus, like the Celts, may have equated all northern peoples with the Germans; or, alternatively, he may have been prejudiced. The full territorial extent of Caledonia is not known; but clearly it was that tract of country falling within Piggott's north-eastern province, in his scheme for recognising four major areas in the Scottish Iron Age.[1]

In battle, these people were no match for the Romans. Agricola won the battle of Mons Graupius in AD 84, somewhere in the heart of Caledonia; and again in 209, the Romans reduced the Caledonians to unconditional surrender. Both these catastrophes must have provoked a spirit of revenge; and it may have been, and almost certainly was, these people who took advantage of Roman weakness to sack and pillage the lands south of the Forth-Clyde isthmus, even to the extent of destroying the native workshops as far south as York. This was the disaster of AD 196.

Timber-laced forts are often said to have been associated with these north-eastern tribes, so that perhaps it is possible to determine the extent of Caledonian territories by the existence of this form of fortification. The numbers are low in the north-east, but increase considerably in the north-west of the province, possibly as a barrier between themselves and the broch builders.[2] Other concentrations exist in Angus and Fife, perhaps with an eye to keeping out the Votadini. There are vitrified forts along the Great Glen, and they overspill into the Firth of Clyde.[3] Whether or not these particular forts had anything to do with the Caledonians cannot be determined; but it must be remembered that they maintained a back-door route to Ireland. Inside the vitrified fort-girt area there is nothing but two centrally placed large hill-top forts, of Dunnideer, near Insch, and Tap o' Noth, near Rhynie. These are fairly large forts, and both are vitrified. There was a sizeable population at Tap o' Noth, judged by the number of hut circles still visible. Perhaps both were oppida, the seats of power in Caledonia.

Had the Caledonians kept their peace, nothing would have been heard of them. They were war-like, and they were in touch with the heroic society in Ireland, perhaps travelling by the old Group IX stone-axe route, which went via Glen Dochart to Ardlui, and thence to the Firth of Clyde and the Mull of Kintyre, before crossing the sea to Ireland. This route was never breached by the Romans. The maintenance of contact helps to explain the eruption of massive armlet decoration on craft-made products all of which have regional characteristics, differing from all others in Britain. However, leaving aside for the moment this suddenly erupting style, the best that can be mustered in matters of art is a large collection of decorated stone balls.[4] Generally, these are sporadic finds, and they occur in most parts of the province, but with a marked concentration along Donside and in Buchan. This is the non-mountainous area of the north-east, which today is mostly agricultural land, and most of today's population lives here. Piggott[5] is of the opinion that some at least of these decorated stone balls are of Neolithic date, though there is a specimen from Fordoun, Kincardineshire,[6] which bears chevron decoration, which is also common on the armlets; whilst in far away Walston, Lanarkshire[7] there is one in cast bronze. If art is skill applied to imitation and design, then there is very little to show in Caledonia.

It is not that the Caledonians lacked a competent metalworking tradition. The number of late Bronze Age moulds for the manufacture of swords and axes, found in this territory, is testimony to a well organised and efficient metalworking industry in these earlier times. But there could never have been any regular contact with the sources of la Tène art: either this was denied to the Caledonians, or they never pursued it. As an exercise, it is interesting to see what the province has yielded in the matter of first- and second-century art objects; and leaving aside, for the moment, massive armlets and Donside terrets, this is what we have:

1. Strap Junction, Inchtuthil, dated to AD 86–90
2. The Drumashie belt-mount and dress-fastener
3. Plain terret and bar, Ardoch
4. Sword, Fendoch, dated to AD 86–88
5. Openwork fragment, Culbin Sands
6. Door-knob type spear-butt, Inverurie
7. Openwork triskele, Clova
8. Finger ring, openwork triskele decoration, Forfar
9. Similar, but in red and yellow enamel, Dunning.

Of the nine objects listed, two were probably brought north by Roman soldiers. There are also imports not listed above: the Deskford boar's head,[8] and the Norries Law[9] plaque and handpins, together with another handpin from Gaulcross.[10] These imports are of Irish origin. The Deskford boar's head and the Norries Law and Gaulcross finds are evidence of a continuing Irish connection, which began in Neolithic times, and which was responsible for the occurrence at Newry, Co. Down, of a massive armlet.

Part of the decoration on the Deskford boar's head consists of a whole concourse of trumpets (Fig. 9:4), all of which are of the elongated form, and typical of Irish decorative work: and together they enclose the eyes. The eyelids are represented by raised quadrants in repoussé style, again a feature of Irish decoration. Piggott's ascription of this boar's head to the middle of the first century AD cannot be faulted. In other words, it is of the same period as are the so-called sacrificial dishes, and before the changeover to the style of decoration represented on the Cork Horns, the Petrie Crown and the Bann disc. The Norries Law silver plaque is of even date with the sacrificial dishes: the elongated trumpets, and the heavily embossed 'twirls' being common to both.

The Irish connection continued for some time after this, as witness the ogham inscribed stones[11] in the province. Yet there is nothing in Irish art that could have given rise to the sudden eruption that brought into existence the massive armlets and the Donside terrets.

Both must be indigenous. These armlets are penannular, with expanded rounded terminals. As long ago as 1880, J.A. Smith[12] divided the known specimens into two forms: oval and folded. The oval form is by far the more numerous, there being 13 of these. In outline, they are symmetrical, and the same can be said for the decoration. In the case of Smith's folded type the form could have been suggested by the snake armlets,[13] of which four examples are known in Caledonia. On the massive armlets the coils are represented by continuous strands, necessary because the armlets are penannular. In all probability, massive armlets were cast flat, afterwards being bent into their final penannular form.

By comparing decoration with the size and weight of the armlets, it must be seen to be effete. This is understandable: since the Caledonians had no art tradition of their own, they had to borrow from the Irish, and they borrowed patterns which happened to be about at that time. They chose the chevron pattern and the elongated trumpet. But they borrowed at a time when repoussé work was already outdated, so what they got was something similar to decoration on the Cork Horns and the Petrie Crown, which, even in Irish art, was an effete form of decoration. Good examples of elongated trumpets can be seen on the Belhelvia armlet (Fig. 9:2). Two other features borrowed from the Irish are also noticeable: these are the broken-back scroll and the raised quadrant, the former occurring on the Loughcrew trial pieces and the latter occurring on the sacrificial dishes. The chevron pattern is a surprise; but then it had found its way into Irish art. On the massive armlets it is used to break up the decoration into zones. However, the repetition of motifs, never properly understood, leads to their becoming rapidly devolved; so that trumpets are replaced by plain scrolls, as on the armlet from Auchenbadie, Alvah, Banff (Fig. 9:3). These scrolls stalk their way over most of the surface of this armlet, repeating the overall pattern pioneered by the trumpets on the Belhelvie armlet. The liberal use of the chevron pattern is a little surprising. Although in use in Ireland, it never became popular there. Instead, it is the kind of pattern that was more common on objects of Roman or Provincial Roman origin,[14] and it is felt that the Caledonians may have acquired it from one such object. Some of the massive armlets have small amounts of enamel. It appears only on small plates inserted into the circular voids in the middle of the terminals. Three examples that come to mind are of the oval type of armlet, from Castle Newe, Strathdon, Aberdeenshire, close to which a denarius of Nerva (AD 96—8) was found, and the two 'folded'-type armlets from Pitkelloney, Muthill, Perthshire. In the first example, the design is cellular, with alternating red and yellow enamels, and the same two colours occur on the two Pitkelloney specimens, but here the design is quatrefoil. It is possible that other massive armlets were enamelled in the same manner, but the discs have fallen out. The quatrefoil pattern is similar to that on the Holderness bridle-bit, the Seven Sisters terret, and the Middlebie petal-shaped strap junction, as well as Brigantian dragonesque brooches. It would appear that the Pitkelloney quatrefoil patterns were borrowed from Brigantia, and that the craft of enamelling was also learned there, since only one colour of enamel is known in Ireland at this time.

With the exception of the Newry specimen, massive armlets have not been found outside Scotland. The majority has come from within the bounds of Caledonia. Below is a list of the find-spots:

Oval Type		*Folded Type*	
Aboyne, Aberdeenshire	(3)	Bunrannoch, Perthshire	(1)
Alford, Aberdeenshire	(2)	Pitkelloney, Perthshire	(2)

Oval Type (cont.)		*Folded Type (cont.)*	
Belhelvie, Aberdeenshire	(2)	Glamis, Angus	(1)
Castle Newe, Aberdeenshire	(2)	Rogart, Sutherland	(1)
Alvah, Banff	(1)	Seafield Tower, Kinghorn, Fife	(1)
Stanhope, Peeblesshire	(1)		
Stichill, Roxburghshire	(2)		

This division into two types is revealing. With the 'oval' type, there is a tendency for the armlets to occur in pairs; whereas, in the case of the 'folded' type, generally they occur singly. The oval type is very much at home in Aberdeenshire and Banff; whereas the folded type belongs to farther south, in Perth and Angus. The oval type alone found its way to south of the Forth-Clyde isthmus. Everything points to there having been two sources of manufacture, one on Donside, and the other (perhaps) in Perthshire, each having a different tradition. The makers of the oval type appear to have been a little nearer to the source of inspiration for the decoration. Most snake armlets occur in Angus and Perth, which lends weight to the suggestion that folded armlets were derived from that source. Maybe both were made in the same workshop. The Culbin Sands spiral armlet[15] might prove to be a very satisfactory link between the two types, with its slightly expanded rounded terminals and its many trumpet representations amongst the decoration.

It is a long jump from Donside to Peebles and Roxburgh, and one wonders how this happened. In these southern counties the associations of the oval type are interesting. The Stanhope specimen was found with the mount of Fig. 56:8, and a Roman patera belonging to the first half of the second century. The pair of massive armlets from Stichill were found with, or near, the collar of that name (Fig. 19). The association with objects such as these, which are not of Caledonian manufacture, implies pillage. Caledonians may have reached the Lowlands early in the second century, when the Romans were in headlong retreat to south of the Cheviot, and the opportunity for pillage would have been favourable. There was a generation's breathing space before the Romans returned, thus giving Donside adventurers plenty of time to acquire the best pickings in the Tweed basin. There are wider implications involved here, and these will be dealt with at the end of this chapter.

Some corroborative evidence is provided by the 'Donside' terret.[16] Again this is a Caledonian form, developed on Donside, where there is the largest concentration and the earliest form. This is perhaps the best and the most practical form of terret in Britain, for it is entirely functional, and it sits very firmly on the harness. The terret is formed of a plain ring, for the most part of fairly even girth, but expanding suddenly at one point on its circumference, and this expansion is cast hollow. Concealed within this hollow there is a bar for attachment. On the inside, there is a central projection (Fig. 43:2). Piggott[17] would derive this form from a Roman military type. Were that so, then manufacture must have started some time after AD 84, the year of Mons Graupius. The suggestion is that the new terret was copied from a Roman specimen picked up on the field of battle. A further possibility is that the very distinctive form of the Donside terret was derived from a little ring terret, open lipped at the bottom, where there is a bar that is continuous with the circumference, and having an undefined small projection on the inside. This is the little terret from Rhynie[18] (Fig. 43:1). This terret was discovered on land at the foot of Tap O' Noth, on the top of which there is a large vitrified fort and many hut sites. But the design was unsatisfactory, for, with an arc-shaped attachment bar it would have been impossible for the terret to sit firmly on the harness. This point was presumably quickly appreciated, which led to the terret being redesigned, and the result is likely to have been something like Fig. 43:2, which comes from Inverurie.[19] All later forms were slightly elliptical, a shape common to terrets of other types. Whereas the inner projection on the Rhynie specimen was rather formless, the possibilities for decoration here were soon realised; so the Inverurie terret has a very nicely moulded inner projection. This central inner projection comes in for a variety of changes, some of which are shown in Fig. 43:3–8. No implied typological sequence is intended here. The Corbridge, Northumberland[20] terret is the most curious of all, for it has been cast in two halves, which subsequently were riveted together by means of three pins.

The main area of distribution for these terrets is again Caledonia. In addition to those named, other terrets have come from Ballestrade, Cromar;[21] Clova;[22] Culbin Sands;[23] Culsalmond; Kirriemuir;[24] and Towie,[25] where a pair was found. No terrets have been found in Fife; but south of the Forth-Clyde isthmus the following have been noted: Cairngryfe, Lanarkshire;[26] Eyemouth, Berwickshire;[27] Oxnam, Roxburghshire.[28] This Caledonian product earned the distinction of having crossed the Scottish border. Specimens have come from Chesters, Northumberland;[29] Corbridge;[30] Giggleswick, West Riding of Yorkshire;[31] Dinas Emrys, Caernarvonshire;[32] Llandrygarn, Anglesey;[33] Billing, Northamptonshire;[34] Linton Heath, Cambridgeshire;[35] and Moorgate, London.[36] There is one from Shetland.[37]

Discoveries of Donside terrets south of the Forth-Clyde isthmus raise some interesting problems. One is the question of date. Only four terrets have associations. The Crichie, Inverurie, terret was part of a small hoard containing a bronze doorknob ferrule and several jet objects, which are probably the heads of pins.[38] They are of a type found in the second- and third-century levels at Traprain Law. The terrets found on the farm of Hillock Head, Towie, were found with several other bronze objects in a cairn, but the terrets alone were preserved. The terret from the Roman supply base of Corstopitum is unfortunately unstratified. So the evidence remains rather negative. Stevenson[39] rightly rejects Alcock's Dark Age proposal; whilst Piggott's suggestion that these terrets mark the trail of the Caledonian alliance, whilst attractive in itself, is nevertheless not in keeping with his own dating of the Torwoodlee broch occupation, which he places as early as the period AD 100–30[40] though this date has been questioned by Stevenson.

Another problem for which there is no ready answer is the possible association of massive armlets with Donside terrets. No such association is known; yet it is possible that both the terrets and the armlets reached the Tweed basin simultaneously. Both may have originated in the same workshop. This would put production into the latter part of the first century, or the beginning of the second. Thus, Piggott's suggestion of a trail left by the Caledonian alliance is too late; but his dating of the Torwoodlee broch is in keeping with the suggestion that massive armlets, Donside terrets, and brochs (or the ability to design and build them) were all part of a contemporary introduction into the Tweed basin from the north-east of Scotland, and that this was at a time when the northern tribes had free access to undefended country south of the Forth, after the headlong Roman retreat to south of the Cheviot. In all the subsequent confusion, a landing would have been possible at Eyemouth, thus avoiding open conflict with the Votadini at Traprain Law, though we know that this oppidum was overrun as well. The massif of the

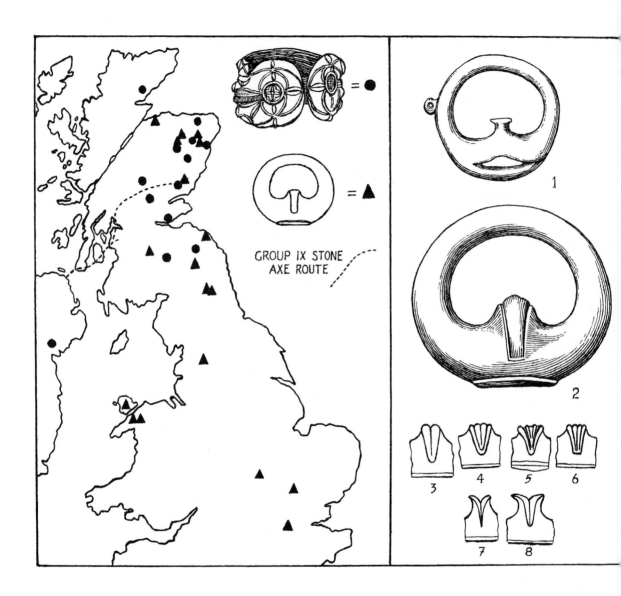

Fig. 43: Distribution Map of Massive Armlets and Donside Terrets *Donside Terrets*: 1. Rhynie, Aberdeenshire. 2. Crichie, Inverurie, Aberdeenshire. 3. Culsalmond, Aberdeenshire. 4. Towie, Aberdeenshire. 5. Corstopitum, Northumberland. 6. Giggleswick, Yorkshire. 7. Kirriemuir, Angus. 8. Oxnam, Roxburghshire. 1 and 2 (3/4); remainder (3/8)

Lammermuirs was between the Votadini and the Caledonians in the Tweed basin, and there the latter remained probably until 142, when the whole frontier system was moved northwards from the Tyne-Solway isthmus, thus bringing the Lowland tribes once more within the Roman aegis. These assumptions may or may not explain why vitrified forts were built in the Tweed basin. The presence of brochs suggests an alliance of the Caledonians with the Caithness broch builders in a military expedition designed to rid the country of the Romans. With such determination, settlement in the Lowlands must have been easy, and for two reasons: first, the Lowland peoples may have received these northern peoples as saviours; and secondly, they had probably been disarmed by the Romans in any case, and would have been in no position to keep out invaders. The new situation left open the road to Wales, making it possible for Donside terrets to get to Caernarvon and Anglesey. It should be noted that these were found on native sites. What is surprising is the occurrence of Donside terrets in England, and as far south as London. There is no ready explanation for their presence here.

After so spectacular an eruption of artistic activity in Caledonia, the fires died down, massive armlets and Donside terrets faded out, and they left nothing behind them. It was almost as though they had never been, and production ceased as mysteriously as it had begun.

NOTES

1. *Roy. Comm. Anc. and Hist. Monuments, Roxburgh* (1956), I, p. 15.

2. Orkney was added to the Roman Empire (Tacitus, *Agricola*, X) but nothing is known regarding the broch builders of Caithness.

3. See the excellent map by I.G. Scott in Rivet (ed.), *The Iron Age in North Britain*.

4. In Rivet[3], Fig. 7 for their distribution.

5. S. Piggott, *Neolithic Cultures of the British Isles* (Cambridge University Press, 1954), p. 332.

6. Anderson, *Scotland in Pagan Times*, Fig. 145.

7. Ibid., 162, Fig. 141.

8. S. Piggott, 'The Carnyx in Early Iron Age Britain', *Antiq. Journ.*, XXXIX (1959), p. 19.

9. Wilson, *Prehistoric Annals of Scotland* (1863), p. 518 and illustration; Macgregor, *Early Celtic Art in North Britain*, no. 349.

10. R.B.K. Stevenson and J. Emery, 'The Gaulcross Hoard of Pictish Silver', *Proc. Soc. Antiq. Scot.*, XCVII (1963–4), pp. 206–9.

11. C. Fox, *The Personality of Britain*, Fig. 18. See also Leslie Alcock's Map in D. Moore (ed.), *The Irish Sea Province in Archaeology and History* (Cambrian Archaeological Association, 1967), pp. 55–65.

12. J.A. Smith, 'Notice of a Massive Armlet . . . from Stanhope, Peeblesshire', *Proc. Soc. Antiq. Scot.*, XV, pp. 316–63. Also Ibid., XVII (1882–3), pp. 90–2.

13. One was actually found in association with a massive armlet; see Smith[12], pp. 337–40, Figs. 17, 18.

14. E.g., the Snailwell bowl. Lethbridge, 'Burial of an Iron Age Warrior', p. 33, pl. VII.

15. Good illustration in Macgregor[9], no. 214.

16. H.E. Kilbride-Jones, 'An Aberdeenshire Iron Age Miscellany: Bronze Terret from Rhynie, and distribution of the Type', *Proc. Soc. Antiq. Scot.*, LXIX (1934–5), p. 448.

17. S. Piggott in C.J. Wainwright (ed.), *The Problem of the Picts* (Nelson, 1955), p. 63.

18. The circumstances of discovery are as follows: this terret was found by the author on Alex Shands' dresser in his living-room. He stated that he had picked it up near the foot of Tap O' Noth. The illustration in Kilbride-Jones[16] is correctly drawn.

19. *BM Guide to Early Iron Age Antiquities*, p. 158, Fig. 189.

20. R.H. Forster and W.H. Knowles, 'Corstopitum: Report on the Excavations in 1910', *Arch. Ael.*[3], VII (1911), p. 188, pl. IV:1.

21. J. Graham Callander, 'Early Iron Age Hoard from Crichie', *Proc. Soc. Antiq. Scot.*, LXI (1926–7), p. 244, Figs. 3 and 4.

22. Ibid., p. 246.

23. Ibid., p. 244.

24. Ibid., p. 246.

25. Ibid., p. 246.

26. V. Gordon Childe, 'Examination of the Prehistoric Fort on Cairngryfe Hill, near Lanark', *Proc. Soc. Antiq. Scot.*, LXXV (1940–1), p. 218, Fig. 2:3 and pl. LII.

27. *Roy. Comm. Anc. and Hist. Monuments, Berwickshire* (1915), XXXVIII.

28. Graham Callander[21], p. 246.

29. Wallis Budge, *An Account of the Roman Antiquities Preserved in the Museum at Chesters*, third plate following page 120.

30. Forster and Knowles[20], pl. IV:1.

31. F. Villy, 'Note of a Bronze Object found at

Gigglewick', *Yorks. Arch. Journ.*, XXII (1913), p. 237.

32. *Nat. Museum of Wales Guide to Coll. illustrating the Prehistory of Wales*, Fig. 44.

33. Macgregor[9], I, p. 71.

34. Leeds, *Celtic Ornament*, p. 126.

35. R.G. Neville, 'An Anglo-Saxon Cemetery excavated in Jan. 1853', *Arch. Journ.*, XI (1854), p. 99, no. XI.

36. Macgregor[9], I, p. 71.

37. Ibid., II, no. 124.

38. Graham Callander[21], p. 243; *BM Guide*[19], p. 158.

39. Stevenson, 'Metalwork and other Objects in Scotland' in Rivet[3], p. 33.

40. S. Piggott, 'Excavations in the Broch and Hill-Fort of Torwoodlee, Selkirkshire', *Proc. Soc. Antiq. Scot.*, LXXXV (1950–1), pp. 92 ff.

18. Dress-fasteners

Increasingly, we shall be dealing with small objects, the sort of objects which the metal-workers made to keep themselves in business when they were no longer permitted to make swords and sword scabbards, spears and shields. Apparently, horse trappings must have been on the list of proscribed items, since, except for a few northern examples, manufacture of these seemingly ceased as well. Great credit is therefore due to the metal-workers who were able to survive the loss of their princely trade, in that they turned to the production of small personal items, such as dress-fasteners, brooches and pins; small compensation, perhaps, but quantity replaced quality. Even so, every effort was made to make these small items attractive to the eye, though that was only at the beginning: with the progress of time, an element of carelessness creeps in, possibly because mass-production techniques were employed.

But, because dress-fasteners, brooches and pins are such small items, they gave the artist little opportunity for exercising his imagination, and even less for artistic expression. So, from this time until the destruction of the workshops, the art becomes soporiferous. Of course, artists may have turned to other mediums, such as wood and leather: we know that some wood carvings have survived. Clearly, the market-place would have been the only outlet for artistic creations, and they are not ideal situations for the disposal of works of art.

Roman togas required no dress-fasteners, so that even in regard to these items, urban populations could not be expected to buy, except out of curiosity. It must be assumed that native countrymen alone would have been the buyers. These dress-fasteners have a button-like appearance, when seen from the front: at the back there is a projecting shank or loop, angled at 90°. The real purpose of these objects has been questioned; but archaeologists generally regard them as dress-fasteners, and as such they are regarded here. The loop at the back was for stitching on to the dress, whilst the button head was inserted in the buttonhole. Throughout, there is little variation in this basic design; but the shape of the head can be divided into six basic styles:

(a) Single and double Boss
(b) Circular
(c) Square
(d) Expanding Spiral
(e) Boss-and-petal
(f) Quoit shaped.

Naturally, there are other forms as well, such as a sort of four-leaved shamrock at Chesters to the odd double-petal fastener from Carlisle;[1] but these are oddities, and their numbers were never sufficient to constitute a style.

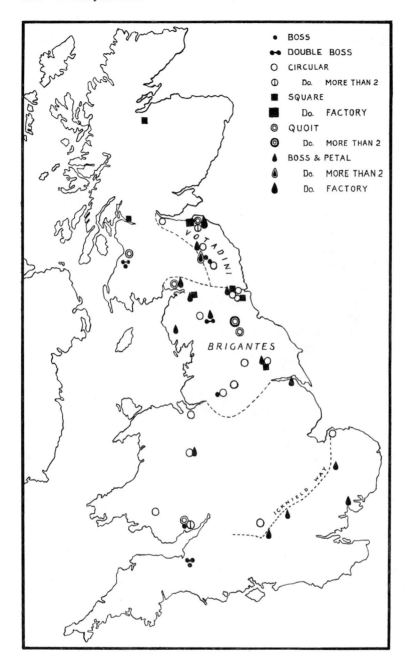

Fig. 44: Distribution Map of Decorated and Enamelled Dress-fasteners.

As befits their purpose, the majority of fasteners have rounded heads, for example, circular, boss-and-petal, expanding spiral, quoit shaped; but, by way of contrast there is the square-headed style, and one or two are rectangular. In the course of his study of these forms, Wild[2] also includes several atypical examples, some seemingly not Celtic. These will not be considered here.

Most dress-fasteners have been found in the former territories of the Votadini and the Brigantes (Map, Fig. 44). There are six strays in the Civil Province, but all are close to the Icknield Way, and the significance of these occurrences will be dealt with later. The trickle to the south of Wales, probably via Chester, is evidence of a possible link-up between the Brigantes and the Welsh tribes, and, in this respect, they followed the route taken by the Donside terrets. So many circular dress-fasteners have been found under conditions that date them by association to Flavian times[3] that their presence in Wales could indicate Brigantian pursuit of the friends of Caratacus, though the middle of the first century might seem a bit early for dress-fasteners. Up till this period, the Romans had not yet penetrated the difficult terrain west and north-west of Gloucester to any real extent. Wroxeter was not in existence before AD 60, but it was left empty in 66. Legio II Augusta, or part of it, made Caerleon its permanent home in 74. But the conquest of Wales was not consolidated until after Agricola's arrival in Britain in the summer of 78. The occurrence of dress-fasteners in a straight line from Chester to Caerleon is a possible indication of the position of the frontier in the days before Agricola's arrival, and his conquest of the hinterland. Alternatively, the items in question might have belonged to Brigantian mercenaries.

(a) Single- and Double-Boss Style

These are prototype fasteners, and both single- and double-boss styles are contemporary. They have been found both with solid and with hollow heads, though hollow heads are more unusual in the single-boss style. Both styles were found at Glastonbury,[4] though here the single-boss fasteners have solid heads. Glastonbury is a later pre-Roman Iron Age site, and as such provides clear evidence of the existence of these two styles in pre-Conquest times. However, there is no other occurrence in the south of double-boss fasteners, the remainder coming from northern sites. One was found at Brough-under-Stainmore, Cumbria,[5] and its association with other objects shows that it is of Flavian date. Another comes from the Lochspouts crannog at Maybole, Ayrshire,[6] but here the bosses are hollow. This is an important occurrence, because it proves the survival of this style into the second century. Along with it was a prototype S-brooch, made of wire, and a Samian bowl of Form D. 37, of second-century date. This dress-fastener is rather more elaborate than others, for there is a centrally placed rectangular panel upon which there is a raised pair of continental trumpet motifs (Fig. 45:12). This decoration puts the dress-fastener into the same period as the cheek-piece from Birrens, Dumfriesshire[7] (Fig. 15:5). This cheek-piece came from Level II on Site VIII, in a homogeneous deposit assignable to the end of a period which closed shortly before AD 158. Other items having the same style of trumpet decoration are the Chesterholm, Northumberland cheek-ring,[8] the ovoid mount from Corbridge,[9] which is either Flavian-Trajanic or Antonine (AD 138–61), and the elongated strap junction from Isurium Brigantium.[10] All these instances show that the style is northern, and that a second-century date is appropriate for a dress-fastener such as the Lochspouts specimen. So that, if the double-boss style began at

Fig. 45: Cast Bronze Dress-fasteners *Single and Double Boss Style*: 4. and 12. Lochspouts Crannog, Maybole, Ayr. *Circular*: 1. York. 2. Ogof-yr-Esgyrn, Brecknock. 3. Masada, Israel. 5. Benwell, Northumberland. 6. Mumrills, Falkirk, Stirling. 7. Slack, Yorkshire. 8. Caerleon, Monmouth. *Square*: 9. Drumashie, Dores, Inverness. 10. York. 11. Cilurnum, Northumberland. (1/1)

Glastonbury, it was then a pre-Invasion style, and somehow it got from here to the far north (perhaps by sea), where it became a northern style current during the early Occupation, and bosses such as these were even used as the sole decoration on a bridle-bit from Middlebie (Fig. 56:1).

The story of the single-boss style is similar. From its beginnings at Glastonbury,[11] it then turns up at High Rochester[12] with Hadrianic associations; at Lochspouts crannog, in an early-second-century context (Fig. 45:4); at Manchester[13] with Flavian associations; whilst another was found at Newstead,[14] but here it is of doubtful date. Once more the north British association is clear, and all northern fasteners have hollow bosses. However, a specimen was found at Caerleon in Room 2 in Barracks VI, and it is dated by association to the period AD 100–20.[15] Also there is an enamelled specimen from the same site (Fig. 45:8), but the dating is anomalous, for, although it was found on a cement floor in Room I in Barrack VI, the occupation debris would date it to the third or fourth centuries.

Otherwise, the history of the single- and double-boss styles run parallel, in that both forms originated in the south-west of Britain, perhaps at Glastonbury itself; but that, after an absence of a century, both reappear as northern styles, and they were in general use in the Flavian-Antonine period.

(b) Circular

Circular dress-fasteners come mainly from Brigantia, and from the territory of the Votadini, with a few from Wales. The numbers from Brigantia exceed all others in the proportion of two to one, so that there can be little doubt but that the type is Brigantian. In Brigantia itself the distribution is eastern rather than western; whilst in Votadinian territory the distribution is mainly along the Roman road from Corstopitum to Traprain Law. In Wales, the distribution is in a straight line from Chester to Caerleon.[16]

In view of their Flavian associations, some circular fasteners might be expected to have a history similar to that of the last style. Nothing could be plainer than a plain disc of metal (Fig. 45:1). Others have concentric channels, clearly intended for enamel. Again, others are decorated with an interesting pattern known as the 'sunburst' pattern, because it is represented as an orb with associated rays. The circular form also becomes the most highly decorated of all types, some showing a maximum use of enamel. Enamelling appears early, and several instances are known with first-century associations. One of the earliest of the decorated fasteners is that shown in Fig. 45:2, which was found in the caves at Ogof-yr-Esgyrn, Brecknock,[17] and it bears traces of blue enamel. First- and second-century trumpet brooches also came from these caves. A more accurately dated specimen (Fig. 45:3) comes from Masada.[18] It is clear that this fastener must have belonged to an auxiliary in the attacking Roman army at the time of the revolt, from AD 66–73, and it can be said to date to the period AD 73–111. An omega brooch was found on the same site. The Masada dress-fastener has a dot of enamel surrounded by two concentric rings, also enamelled. There is another fastener of similar design (Fig. 45:5), except that the decorated disc is here mounted on another, which is both plain and larger. It comes from Benwell, on Hadrian's Wall.[19]

A native dress-fastener, but with a very non-Celtic type of decoration, is that from the Roman fort at Mumrills.[20] Here there was an Agricolan occupation of short duration in AD 80 or 81. The fort did not become important again until about 142, when the attempt was being made to regain some of the ground lost after the first Agricolan occupation. The Samian and coarse ware found here are characteristically the same as those found on Antonine sites in Britain; and for that reason the fastener has been assigned to the Antonine period.[21] However, this is too late for this fastener, for subsequent development of the Sunburst pattern indicates that the Mumrills specimen was a prototype design, in which case it is much more likely to have been lost during the initial phase of occupation in 80 or 81. The disc has a milled edge, a feature noted on some first-century disc-brooches. The decorative pattern is a clever representation of an orb aflame. Superimposed on it at the centre is the Celtic quatrefoil pattern, also current in the first century, and seen on objects like the Holderness bridle-bit and the upper Dulais (Seven Sisters) terret. The Sunburst pattern was first noted on that Provincial Roman object, the Snailwell bowl (Fig. 18:5), which was buried shortly after the Claudian Invasion of AD 43. Main points to note in the case of the Mumrills pattern are the very realistic tongues of flame shooting out from a central orb. The whole pattern is enamelled in alternating colours of red and yellow.

No pattern ever found more immediate favour with the native metalworkers; but the naturalistic representation on the Mumrills fastener was modified to suit native taste, with the result that the pattern was simplified into a more abstract form, as on the specimen from Slack, Yorkshire[22] (Fig. 45:7). Here the enamel is contained in equally spaced cloisons, in which the dividing strips are diamond shaped. The central orb is represented by a circle which encloses an abstract version of the quatrefoil pattern. The enamel is in blue and red. This dress-fastener was found in the fort annexe, and by association it can be dated to *c*. AD 80–140. Finally, the sunburst pattern in its abstract form was stabilised in a representation which consisted of a series of well balanced enamel-filled triangles; but this development was reserved for the umbonate disc-brooches, already discussed (see Fig. 18).

The daisy-like pattern on the Caerleon dress-fastener (Fig. 45:8) is also based on this sunburst pattern: another very similar came from the Roman cemetery on The Mount, York,[23] and yet another was found at Cilurnum.[24]

(c) Square

A complete mould for casting a square dress-fastener was found in the oppidum of Traprain Law, East Lothian (Fig. 46:1). This is proof of local manufacture. Square dress-fasteners have been found in the lowest levels here, thus dating them by association with datable objects to the second half of the first century AD. The majority are plain, clumsy, and the workmanship is inferior, perhaps because they were turned out in their hundreds in the Traprain workshops. If decoration occurs, it is unimaginative; nothing more than a saltire (Fig. 46:4) or an M, as at Chesters. When enamelling occurs, the enamel is contained in squares and triangles.

Confirmation of the first-century Traprain Law date is provided by the circumstances of discovery of a specimen in the bath house at The Red House, Beaufort, near Corbridge, Northumberland.[25] Here the fastener was associated with Samian ware, all of first century manufacture, with deposition dated to *c*. AD 90. The Beaufort building was dismantled shortly after the Flavian fort at Corbridge was sacked in *c*. AD 98. This Red House dress-fastener is fairly large, being an inch square, though the largest is a plain specimen found in a metalworker's shop at Stanwix, Carlisle.[26] Stanwix could have been occupied as early

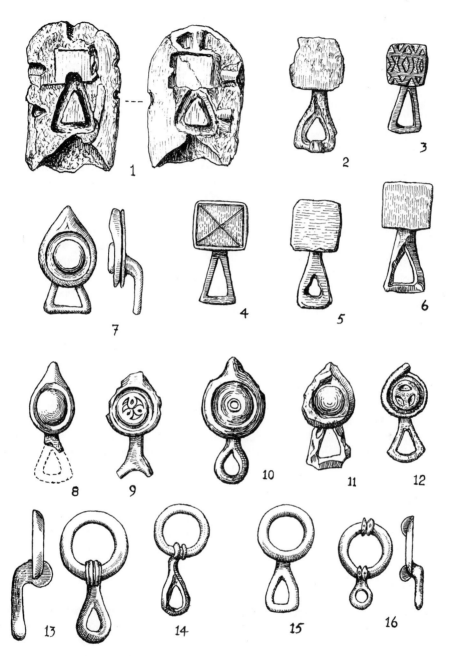

Fig. 46: Cast Bronze Dress-fasteners *Square*: Clay Mould for Casting Dress-fastener, Traprain Law.
2.–6. Traprain Law. *Expanding Spiral Form*: 12. Traprain Law. *Boss-and-Petal*: 7. Wall Area.
8.–11. Traprain Law. *Quoit-shaped*: 13. Caerleon, Monmouth. 14. 15. Traprain Law. 16. Stanwick,
Yorkshire. (3/4)

as the building of Hadrian's Wall in 122, and occupation had ceased by *c*. AD 140. Square-headed dress-fasteners persist through all levels at Traprain Law, from the lowest to the top level, thus demonstrating the persistence of this fashion, even after the disaster of 196. Such instances of the revival of a form, after the abandonment of the workshops for some 30 or 40 years, and at the original place of manufacture, are rare. This information was not available to Reginald Smith when he wrote that the square fastener dated to *c*. AD 100.[27]

Plain, square dress-fasteners were obviously intended for everyday use, in everyday wear, and there are a lot of them. Less numerous are the enamelled specimens. Still somewhat crude and badly finished, they look as though they too were made in a hurry. The fastener (Fig. 45:9) from Drumashie, Dores, Inverness-shire[28] is the only specimen ever to have been found north of the Forth-Clyde isthmus. Brigantian metalworkers were capable of producing much better finished products, as may be seen by reference to the dress-fastener from York (Fig. 45:10). Here the finish is good. It is decorated with isosceles triangles enamelled in alternating colours, with a panel of squares across the centre. A reversion to full Celtic style is recorded on the rectangular dress-fastener (Fig. 45:11) from Cilurnum.[29] The ornamentation is similar to others in linear style on the St Keverne mirror and other mirror-style designs. The Cilurnum fastener cannot be earlier than 122, since this fort was a secondary addition to the Wall system.

(d) Expanding Spiral Form

This is a rare but interesting form, and it is mentioned here solely for the reason that it is intermediary, leading directly to the fully developed boss-and-petal style. Basically, it consists of a trumpet wrapped around a central non-boss (Fig. 46:12); this specimen comes from Traprain Law. Non-bosses have flattened tops, which are often decorated. Here the decoration consists of a spherical triangle on an enamel background. The resemblance to the trumpet enclosed decorated non-bosses on the Holderness bridle-bit (Fig. 26:1) is very close indeed. It is easy to appreciate how the boss-and-petal evolved from this style.

Everything points to this dress-fastener being of first century date: style, the presence of the spherical triangle are but two pointers; so that it comes as a shock to find that a pair of these expanding spiral dress-fasteners came from the top level at Traprain Law. They are completely out of context here, and one must reserve judgement on this occurrence.

(e) Boss-and-Petal

This appears to have been the most popular form of dress-fastener, and it was turned out in thousands. Basically, the design consists of a central boss, hemi-spherical in shape, and set in a hollow in the middle of a petal, so called because it is not circular, but is extended to a blunt point on one side only.[30] Occasionally, the boss is slightly flattened on top for the purpose of receiving some rudimentary decoration. As a style it represents the ultimate in the development of the expanding spiral form. Sometimes there is a medial line on the point of the petal, which was intended for the purpose of indicating the former presence of the trumpet mouth.[31] All the indications are that this style was evolved at Traprain Law, with some help from outside.

The first-century boss-style vogue led to the development of the boss-and-petal; and the Traprain Law evidence suggests that boss-and-petal fasteners are as early as the first century (bottom level), but that they occur also in the third level. James Curle thought that bosses allied to petals, as a style, were current at least by AD 100.[32] Leeds managed to distinguish a northern boss style, which he claimed was characteristic of the eastern Lowlands and southern Scotland.[33] This theory would bring the whole style within Votadinian territory, which the map, Fig. 44, confirms. Perhaps the sole manufacturing source of these fasteners was Traprain Law, which promoted a healthy trade with Brigantia. There is an interesting overspill into the Civil Province which will be discussed later.

Gillam dated these dress-fasteners to the Flavian period.[34] They have not been found in any of the Wall forts, which is a significant fact, for on these grounds it would appear that manufacture had ceased by 122. Yet, at Traprain Law itself some were found in the third level, or in the 'native phase' which occurred here after the abandonment in 196. But the occurrence of two fasteners in the Middlebie hoard places them in the first, rather than in the second century. Unfortunately, Newstead adds no further information regarding date. In the southern overspill, the Lowbury Hill specimen[35] was unstratified and even the Colchester fastener[36] (Fig. 26:3) had no datable associations.

Only a total of four boss-and-petal fasteners are enamelled. One (Fig. 46:10) has a spot of enamel at the centre of the boss, whilst Fig. 46:9 has a trefoil pattern occupying the same position. The fasteners from South Ferriby (now lost) and York were both enamelled in blue.

(f) Quoit-shaped

Although there are not many of this type, nevertheless they are widely distributed, having been found at Traprain Law, in Brigantia, and at Caerleon in south Wales.

They are well-named.[37] The style is definitely Brigantian. No less than 13 fasteners were found in the Stanwick hoard,[38] the largest single collection, and all of them can be dated to *c*. AD 50–60. Thus this style is one of the earliest. As with so much of the Stanwick metalwork, these quoit-shaped fasteners are lip-ornamented (Fig. 46:16), and in some cases there are enamel settings; one is gilt. The lip mouldings are an eastern British (Brigantian) style, though Brailsford would prefer a derivation from what he calls the Polden Hill style,[39] where massive lip-mouldings are numerous on terrets. The Polden Hill lip-mouldings are extremes of the eastern British forms, in which area they are restrained, like those at Stanwick itself.

Beyond the borders of Brigantia, quoit-shaped dress-fasteners have been found at the Lochlee crannog, Ayrshire,[40] and at Caerleon, Monmouth,[41] at which site they are contemporary with the first occupation of Barrack IV, and here they were associated with pottery dating to AD 80–120. The Middlebie specimen came from Brigantian territory, and it was associated with the hoard of that name, and which is said to be of late first-century date. But the two specimens found at Traprain Law came from the top and second levels, implying that they belong to the 'native' levels. This places them in the post-196 years. The fastener from the second level (Fig. 46:14) is a very primitive looking effort, with the top of the loop merely twisted round the ring and then lead-soldered. This is not design work, but the work of a copyist. But the second, from the top level, is a more professionally made fastener (Fig. 46:15), with ring and loop in a single casting. The maker of the Caerleon dress-fastener (Fig. 46:13) may have been Brigantian, but he shunned lip-ornament. Lastly, a fastener found at Catterick[42] was not stratified.

Dress-Fasteners found in the Civil Province

The Civil Province was that non-military region situated to the south of Brigantia. Until AD 74, Brigantia had existed as a buffer state between this Province and the northern tribes. But after Brigantia was brought to its knees in 74 by the Romans, in retribution for Venutius' misdeeds, it then became a policed state, and the military zone was taken farther north. Because of the proselyting activities of the Romans, everywhere there was a growing absence of continuing Occupation styles, especially in the second century, implying that manufacture was continuing only in the fringe areas. A reversal of feelings on the part of the Britons might be thought of as encouraging urbanisation, in which ancient oppida became towns, whilst other towns were established on other well chosen sites, between which interconnecting roads were built. In the civitas-capitals traders and manufacturers contributed to their growth. The tribal aristocrats now had their country estates, but no longer was it possible for them to engage in private warfare. Wealth could be increased by investing in trade, or in manufacturing industries, or even in urban property.[43] But in spite of the benefits to be derived from the spread of urbanisation, and the current ease of travel, there must have been many who hankered after the old ways, as people still do today. By nature, the Britons were not town dwellers.

It would appear that the old Icknield Way, a highway since Neolithic times, remained in use, preferred perhaps, to the new road system. It begins at the coast, somewhere near Ringstead, and it snakes across country well into the upper Thames Valley. Either the Romans turned a blind eye to its continued use, or else they were unaware of its existence. The distribution map, Fig. 44, shows that it was being put to good use by merchants selling time-honoured native-made products, which included dress-fasteners. A short sea-trip from the mouth of the Humber would have brought them to the terminal of the Icknield Way; and at Ringstead our first dress-fastener, a circular one, was found.[44] It was found in a hoard of Iron Age metalwork. Another circular dress-fastener was found at Campsfield, Oxford,[45] but still close to the Icknield Way. The rest of the fasteners found are in the boss-and-petal style, made at Traprain Law. One was found at Hockwold, Norfolk,[46] in an Antonine context; another came from the Roman villa in Gadebridge Park, Hemel Hempstead, Hertfordshire,[47] and here dated to the middle of the second century; whilst a third was found on Lowbury Hill, Berkshire,[48] an Iron Age and Romano-British site, but unfortunately the fastener was unstratified. The last specimen comes from Colchester,[49] which is some way from the Icknield Way, but nevertheless it could have come by that route.

Was this trade conducted by subterfuge? Perhaps the roads were barred to native Brigantian traders selling trinkets made in the fringe areas. Or did the manner of distribution follow an age-old pattern? We shall never know; but the fact remains that dress-fasteners were still saleable in a Province where the toga had probably superseded native dress; and that those who bought these articles were Britons wishing to maintain the old ways of life and dress.

NOTES

1. F. Henry, 'Emailleurs d'Occident', *Prèhistoire*, II (1933), p. 88, Fig. 11:3.

2. J.P. Wild, 'Button and Loop Fasteners in the Roman Provinces', *Britannia*, I (1970), pp. 137 ff. Wild's

division of dress-fasteners is fundamentally unsatisfactory, in that no clear pattern of distribution emerges. Equally unsatisfactory is the division into types by numbers.

3. Gillam in Richmond (ed.), *Roman and Native in North Britain* (1958), pp. 79–85.

4. Bulleid and St George Gray, *The Glastonbury Lake Village*, I (1911), p. 219, pl. XLII, E.159, E.174.

5. Gillam[3], pp. 80, 90.2; Macgregor, *Early Celtic Art in North Britain*, I, Fig. 7:10.

6. Munro, *Lake Dwellings of Europe*, p. 423, Fig. 153.

7. Eric Birley *et al.*, 'Excavations at Birrens', p. 337, Fig. 38:3.

8. *Proc. Soc. Antiq. Newcastle*[2], IV (1891–2), p. 269, Fig. 2.

9. Macgregor[5], no. 15.

10. Ibid., no. 30.

11. Bulleid and St George Gray[4], p. 219, E.151.

12. *A History of the County of Northumberland*, XV (Andrew Reid, 1940), p. 154, no. 3.

13. F.A. Burton, *The Roman Fort at Manchester* (Manchester University Press, 1909), p. 158, pl. 92.

14. Curle, *A Roman Frontier Post and its People* (1911), pl. XCII:5.

15. A. Fox, 'The Legionary Fortress at Caerleon, Monmouth', *Arch. Camb.*, XCV (1940), p. 130.

16. It was also the frontier marked by a concentration of native hill-forts, and stretching in a straight line from the Dee to the Severn. See A.H.A. Hogg in *CBA Research Report*, no. 9, p. 12, Fig. 1.

17. G.C. Boon, 'The Roman Material from Ogof-yr-Esgyrn, Brecknock', *Arch. Camb.*, CXVII (1968), p. 49, Fig. 11:7. It is listed as a strap terminal.

18. Y. Yadin, *Masada* (Sphere, 1973), p. 150, top right where it is listed as a knob.

19. J.A. Petch, 'Excavations at Benwell (Condercum)', *Arch. Ael.*[4], IV (1927), p. 189, pl. XL:2, 16.

20. George Macdonald, 'The Roman Fort at Mumrills', *Proc. Soc. Antiq. Scot.*, LXIII (1928–9), p. 555, Fig. 115:13.

21. Wild[2], p. 151.

22. Gillam[3], p. 81. Also *Yorks. Arch. Journ.*, I (1869–70), p. 11 and Fig.

23. Dickenson and Wenham, 'Discoveries in the Roman Cemetery on the Mount, York', *Yorks. Arch. Journ.*, XXXIX (1958), p. 317, Fig 13:89.

24. Wallis Budge, *An Account of the Roman Antiquities Preserved in the Museum at Chesters*, no. 1059.

25. C.M. Daniels, 'The Roman Bath House at Red House, Beaufort, Corbridge', *Arch. Ael.*[4], XXXVII (1959), p. 156.

26. R.G. Collingwood, 'Roman Objects from Stanwix and Thatcham', *Antiq. Journ.*, XI (1931), p. 41, Fig. 1:8.

27. *BM Guide to Early Iron Age Antiquities*, p. 151.

28. *Proc. Soc. Antiq. Scot.*, LVIII (1923–4), p. 11, Fig. 2.

29. Chesters Museum, no. 2941. Budge[24], no. 1396.

30. Wild[2], calls this form 'the teardrop style'. He completely overlooked the significance of the form, and how it came about.

31. Clearly visible on the Colchester fastener; *BM Guide*[27], Fig. 177.

32. In the *Journ. Roman Studies*, III (1913), pp. 99ff.

33. Leeds, *Celtic Ornament*, p. 111.

34. Gillam[3], pp. 80–1.

35. D. Atkinson, *The Romano-British Site on Lowbury Hill* (University College, Reading, 1916), pl. XIII:6.

36. J.H. Pollexfen, 'Antiquities found at Colchester', *Arch.*, XXXIX (1863), p. 508, pl. XXIV:10.

37. 'Quoit' as a term is more apt than 'ring-headed'. A quoit is a flattish sharp-edged ring, a description that exactly fits these fasteners.

38. M. Macgregor (nee Simpson), 'The Early Iron Age Metalwork Hoard from Stanwick, N.R. Yorkshire', *Proc. Preh. Soc.*, XXVIII (1962), p. 17, Fig. 8:23, 24, 25, 26, 28, 30, 31, 32, 33, 34, 35, 36.

39. Brailsford, 'The Polden Hill Hoard, Somerset', p. 222, pl. XVII.

40. Munro, *Ancient Scottish Lake Dwellings*, p. 132, Fig. 147.

41. Fox[15], p. 134, Fig. 7:31.

42. *Yorks. Arch. Journ.*, XXXVII (1950), p. 418.

43. S.S. Frere, *Britannia*, p. 275.

44. *Proc. Preh. Soc.*, XVII (1951), p. 222, pl. XIX:2.

45. *Journ. Roman Studies*, XL (1950), p. 100, Fig. 19:8.

46. *Proc. Leeds Phil. and Lit. Soc.*, XII, part II, Fig. 13:2.

47. *Soc. Antiq. Research Reports*, XXXI (1974), p. 129.

48. Atkinson[35], pl. XIII:6.

49. Pollexfen[36], pl. XXIV:10.

19. Dragonesque Brooches

In the matter of its basic form, this brooch resembles a capital S, with terminations in the form of heads at the top and bottom extremities. The heads serve to keep the pin captive. The final form has been called 'dragonesque' because, mistakenly, it was assumed to have hippocampic qualities. The type is Brigantian, designed and manufactured in that country.

The original manufacturing locality was remote from all others. Collingwood[1] thought that the progenitor of the series had been a little brooch from Braughing, Hertfordshire. Where he led, others followed. He had good reason for his selection, since the Braughing brooch is early, very early in fact, as it was associated with other objects like thistle brooches of a type found on the Continent on Augustan sites, thus suggesting a date only slightly later than the year of the Claudian Invasion of Britain.

But, as a progenitor, a better claim can be put forward on behalf of the simple S-brooch. The earliest form is nothing more than a bent piece of wire, with coiled ends (Fig. 48:1). The coil at one end is always bigger than is the coil at the other. Three of these crude little brooches were found in the Victoria Cave, Settle,[2] and another in nearby Sewell's Cave.[3] A fifth specimen came from the Lochspouts crannog, Maybole, Ayrshire,[4] and a sixth was found at Newstead, Roxburghshire.[5] These brooches are so unsophisticated that anybody with a piece of wire and a pair of tongs would be able to make one. Their association with the Victoria Cave is significant, since the true S-brooch form[6] (Fig. 48:2) was apparently devised here also.

This solid cast S-brooch form is made up of two cornucopia motifs conjoined, mouth to mouth. The design is quite striking, and certainly it did not deserve Collingwood's taunt of being meaningless. The use of these motifs gives meaning to the overall design,[7] and because of their utilisation, a first-century date is suggested. But the new form was short lived. It is not clear what instituted the change, but somewhere a new design emerged. That occurrence must have been in Brigantia, though the earliest brooch to the new design is from Lakenheath, Suffolk[8] (Fig. 48:3). This is an openwork brooch design, carefully thought out, making the finished product both beautiful and sophisticated. But the style is formalised. There is a roundel at the centre, made up of two trumpets in much the same way as the similar though more complex roundel on the Thames at London bridle-bit (Fig. 15:4); so that perhaps the two items are more or less contemporary. On the Lakenheath brooch, and for the first time, are 'heads'; but they are not heads in the strict sense, for they represent no known zoological species. They are, in fact, capped trumpets. But for the 'capping' they could be said to be cornucopias. Each is attached to the brooch midway down its back curve. Nobody can be blamed for adding an 'eye', because, by being put there in the first place, for the purpose of keeping the pin captive, this transverse termination had already created the impression of being a 'head'.

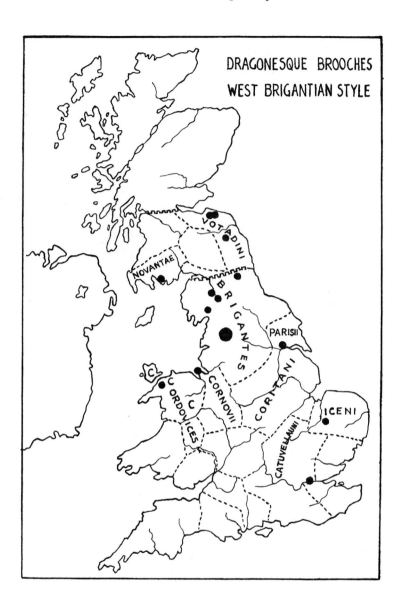

Fig. 47: Dragonesque
Brooches: Distribution
of West Brigantian
Style.

(a) West Brigantian Style

Although the Lakenheath brooch was found in Suffolk, the style is not of those parts. All
early developments took place in Brigantia. A large proportion of the brooches came
from the Settle area, and it is known that metalworking took place in the Victoria Cave;
so that, for these reasons, and because distribution confirms it, some 20 brooches can be
said to be West Brigantian.

All the brooches under this heading are cast. The S-brooch from the Victoria Cave does
not give a fair dating for the initial development, for it was found with brooches of other
forms, all of which are of Flavian date. On the other hand, the Lakenheath brooch, which

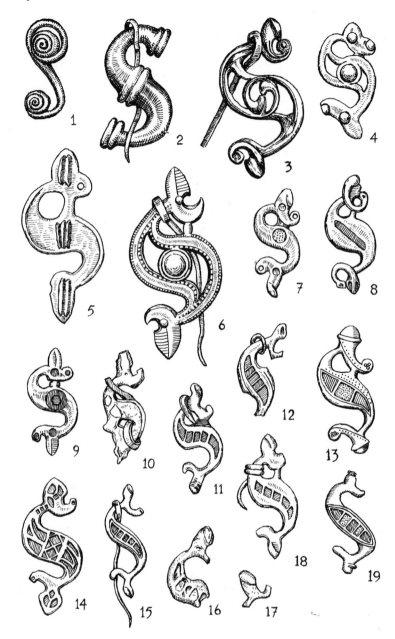

Fig. 48: Dragonesque Brooches, West Brigantian Style 1. 2. 7. 9. Victoria Cave, Settle, Yorkshire.
3. Lakenheath, Suffolk. 4. Edgerston, Jedburgh, Roxburghshire. 5. 6. Attermire Cave, Settle, Yorkshire.
8. Watercrook, Kendal, Cumbria. 10. 16. Traprain Law. 11. Corstopitum, Northumberland.
12. Segontium, Caernarvon. 13. Brough-under-Stainmore, Cumbria. 14. Kirkby Thore, Cumbria; 15.
Tokenhouse Yard, London. 17. Borness Cave, Kirkcudbright. 18. Meols, Hoylake, Cheshire.
19. Foxfleet, Yorkshire. (3/4)

represents the fully fledged dragonesque form, was found in a later pre-Roman Iron Age inhumation, which Rainbird Clarke[9] places in a federate Icenian context of *c*. AD 40–60. Against such a count, the Lakenheath brooch could predate the Braughing specimen, support for which thinking comes from the fact that the Braughing brooch is a plate brooch, of a type still unheard of. For plate brooches postdate openwork cast specimens. The Lakenheath design is clearly a breakthrough, the ultimate in good design work; but ultimates of any form always occasion modification.

So, what followed? All subsequent brooches were either simplified versions, or else there was a switch to the manufacture of plate brooches. But the 'capped' trumpet-shaped head remained with them throughout, though the approaches to shape were various. However, the basic idea behind that shape was never forgotten. The differentiation between openwork and plate brooches is marked. Enamelling is rare on openwork brooches, but quite common on plate brooches. Decoration is transitory, varying very much according to current styles. Eccentricities are common, and they can often be traced to a common source.

Seven brooches were found in the Settle Area. Settle is well placed, being within easy reach of the north Wales copper mines. The Hoylake and Segontium brooches could be pointers to the route followed by those seeking raw copper from the Romans, who took over these mines not long after the Conquest.[10] Copper supplies were also available in Scotland, in the territory of the Novantae,[11] and the Borness Cave brooch might indicate contact here.

Settle was almost certainly a factory area. Whether or not one can be more precise is open to discussion; but the pattern of distribution clearly indicates that both development and production took place in this region, perhaps during the period of the emergent boss style. There is a reason for this statement: for the openwork style of the Lakenheath brooch persisted, but the central roundel was replaced by a central boss (Fig. 48:4). This Edgerston, Jedburgh, brooch has a twin in the Kelco Cave, Settle, establishing a clear relationship between the two areas, the one in Votadinian territory, the other in Brigantia. However, plate brooches were soon to come: an early specimen is Fig. 48:5, from the Attermire Cave, Settle,[12] on which crude raised mouldings are a dim recollection of conjoined cornucopias. The heads are highly stylised, and Collingwood's reference to 'senseless mouldings' perhaps shows a lack of understanding. In contrast, a second Attermire brooch (Fig. 48:6) is expertly made, still keeping its central boss, but the head form is a little odd, being rather pointed in shape. There is a feeling of uncertainty here, which relates more to the heads than it does to the bodies. Head shape is constantly undergoing change, yet the hemispherical cap, of the capped-trumpet form, remains throughout, but sometimes it looks like a curl of hair against the forehead. The most faithful representation of this capped-trumpet form is to be seen in the case of Fig. 48:13, which comes from Brough-under-Stainmore.[13] For comparison, one should look at the capped trumpets of the Drumashie belt-plate (Fig. 15:6).

When brooches become smaller, they also appear to have been mass-produced. Fig. 48:7 is an example: it is not yet a true plate brooch, and it comes from the Victoria Cave, Settle.[14] Central bosses have given way to enamel-filled roundels, both in this and in the case of Fig. 48:9: one colour in the first and two colours in the second brooch. But fashions in decoration were fluid, and a change is coming about in the case of the specimen from Watercrook, Kendal, Cumbria (Fig. 48:8), and the idea becomes finalised in all brooches subsequent to Fig. 48:9. The generally devolved character of the decoration is marked: there are no art motifs, and workmanship is poor, reminding us of Collingwood's meaningful remarks.[15] Clearly, this was the period of the factory-made article.

Not many of these brooches had datable associations. Fig. 48:2 and 3 have mid-first-century-AD associations: Fig. 48:11 is from Corstopitum, and it came from Site XI, where the Samian ware indicates extensive occupation in the second century, from *c.* 120–40 onwards; but there is definite though scanty evidence of a previous occupation dating from Flavian times.[16] The Segontium brooch, Fig. 48:12, is dated to about the year AD 100.[17] The fragment, consisting of a head only, (Fig. 48:17) which came from the Borness Cave, Kirkcudbright, was found amongst a collection of artefacts that included fragments of glass armlets and part of a cup of Form D. 27, all suggesting a first-century occupation.[18] The two Traprain Law specimens (Fig. 48:10 and 16) came from the lowest level, showing that they were contemporary with the period of increased occupation of the oppidum which occurred towards the end of the first century AD. By taking all this information relating to date into consideration, it can safely be said that the West Brigantian-style brooches were being made for about half a century, from the middle to the end of the first century AD.

(b) East Brigantian Style

This style could well be called Parisian. The distribution of the brooches (Fig. 49) is concentrated along the border dividing the Brigantes from the Parisii, so that the choice becomes a difficult one. These brooches avoid west Brigantia, but a few managed to penetrate the Civil Province, and one got as far as Kent. Ultimately, it was decided that the Parisii had a style of their own, which will be considered later.

The first noticeable characteristic of the eastern style is the superior workmanship of the brooches. They have been carefully made, and the finish is good. Everyone is a plate brooch. Almost the whole body is covered with champlevé enamel decoration, and amongst the patterns are a few of Celtic origin. But the method of conjoining heads to bodies differs from the West Brigantian style, in that the neck is joined to the top of the head, and to make up for the imbalance caused, 'ears' were added. Even so, the heads are still basically trumpet shaped, whilst the ears are petal shaped. The addition of ears gives a real zoomorphic character to the brooch as a whole. In this style, the ears have medial strips, dividing the enamel into two arcs. Sometimes this medial strip is waisted, and sometimes it is not. If the ear seems large, it is big of necessity, since there would be a definite imbalance with a smaller ear. This upstanding ear is like a trademark: all the brooches bear such a strong family resemblance that they must be the products of a single workshop.

Almost identical are the brooches of Fig. 50:1, 2 and 3. The first comes from York,[19] the second from Corbridge, but the third has no locality.[20] Also Fig. 51:1 and 2 are identical, though not from the same mould; yet the former comes from Norton, East Riding, Yorkshire,[21] whilst the latter comes from Newstead, Roxburghshire.[22] Any slight variation in decoration is representative of but a passing phase. So far as the majority is concerned, it is clear that multiple lozenges were favoured as a means of decoration. Lozenges are not a native art pattern, which implies that they were borrowed from alien sources, perhaps from Colchester,[23] early in Claudian times.[24] There is a suggested second half of the first century date for these brooches.

There are always strays, and one reached Wroxeter,[25] where it was found under the *opus signinum* floor in Room 5, with several coins, the latest being of Hadrian, and pottery which is not later than *c*. AD 130. Wroxeter was not built before AD 60, and although the coin and pottery evidence might suggest an early second-century date for this brooch, this is unlikely, because nowhere else does the evidence support so late a date. The brooch may have been acquired late, or it may have been an heirloom. The scroll work, incorporating small circles and enamel filling, and best exemplified in Fig. 50:3, can be equated

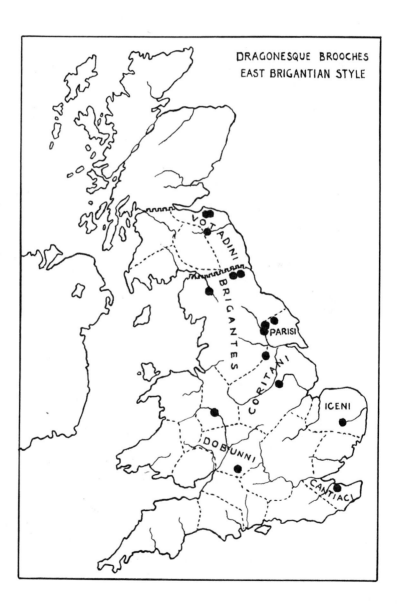

Fig. 49: Dragonesque Brooches: Distribution of East Brigantian Style.

with the decoration, based on the swash N pattern, seen on the Lincoln seal-box lid (Fig. 56:4), and influenced by the style of Fig. 22:3. A date in the second half of the first century AD is most likely for this brooch. On none of the brooches of Fig. 51 is there a suggestion of this form. Also, Fig. 50:1, 2 and 3 have panels decorated with multiple-lozenge patterns, never intended by the Romans to be represented like this, but now recognised as a native adaptation, and transferred in this form to Ireland, where it was included amongst the recorded patterns on the Dooey trial piece (Fig. 67:1a). The lozenges on the Wroxeter brooch (Fig. 50:5) are closer to the Roman originals, being represented like this on head-stud brooches, one of which was also found at Wroxeter. Head-stud brooches are a northern type, belonging to the early second century. The only brooch with this decoration so far unmentioned is Fig. 51:5, from Traprain Law.

Then these rectangular panels with their lozenge decoration are dropped, and the preference is then for central roundels. So far as Fig. 51:6 and 7 are concerned, these include quatrefoil patterns on a background of enamel. The first brooch mentioned came from Margidunum, Nottinghamshire,[26] and the second brooch came from York. Both are strongly similar. The same can be said of Fig. 50:6, from Cantley, Doncaster, and Fig. 51:9 from Kirkby Thore, Cumbria, with their four small squares within a central roundel. These similarities presumably express relationships, and together they throw some light on distributive patterns. The discovery of two brooches (Fig. 51:4 and 5) at Traprain Law suggested local manufacture to one commentator,[27] but this seems improbable. The pair of brooches from Faversham[28] are certainly a long way from Brigantia, and their presence among the Cantiaci is difficult to explain.

There is little to go on with regard to date. The two Traprain Law brooches came from the lowest level, implying a late-first or very-early-second-century date. For the Newstead specimen there is a *terminus ad quem* date of AD 105, the time when the fort went up in flames. Newstead was a Flavian foundation, so that the brooch could have been lost at any time between *c*. AD 70 and 105. There is little more to add, except to say that the East Brigantian style made a later start than did the West Brigantian style, but petered out at virtually the same time.

(c) Parisian Style

This is a very distinctive style. The brooches are heavy in appearance, and are often ill-proportioned. Their most noticeable characteristic is a pointed head, which in some cases is almost cone shaped. The pattern of distribution (Fig. 52) is not unlike that of the East Brigantian style, but now there is a slight concentration in the territory of the Parisii; and for this reason, it is thought that the workshop may have been situated in an area occupied by these people. This style is not represented at Traprain Law, though one brooch was included amongst the scrap in the Lamberton Moor hoard.

Now it would appear that the underlying motif governing head shape has been forgotten at last. Even so, heads still retain something of their former appearance; but here the tendency is for the head to slope backwards, and its top is pointed. Decoration is sterile, lacking any refinement. Normally it consists of little more than square or elongated pointed cloisons filled with enamel. The squares are always grouped together at the centre of the body, clearly taking the place of a panel. The enamel filling is more often than not in two colours, though the colours themselves range from reds through yellows and browns to blues, greens and creams.

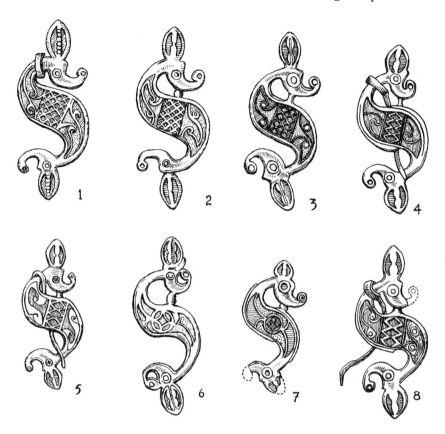

Fig. 50: Dragonesque Brooches, East Brigantian Style 1. York. 2. Corbridge, Northumberland. 3. No Locality. 4. 8. Scole, Norfolk. 5. Wroxeter, Shropshire. 6. Corbridge, Northumberland. 7. Cantley, Doncaster. (3/4)

The brooch from Tanner Row, York[29] (Fig. 53:1) must be one of the earliest: it is clumsy, with a central roundel in two colours of enamel. The inscribed arcs on the body must have suggested later decorative layouts, as at Stanwix, Carlisle[30] (Fig. 53:3) and Milking Gap, Northumberland,[31] whence brooches have central roundels. The Milking Gap brooch (Fig. 53:4) is the best of this series, being slim and nicely finished. Its central roundel is composed of two small circles within a third, a pattern that has survived from much earlier times, and seen on the Thames spearhead (Fig. 32:4) and the Meare scabbard (Fig. 32:2), its origins being traceable to Stamford Hill, Plymouth and Lisnacrogher (sword scabbard, Fig. 36:1). The same pattern occurs frequently on some Glastonbury ware. The pattern is a very ancient one, and it has been discussed by Jacobsthal.[32]

Corstopitum[33] yielded the dragonesque brooch shown in Fig. 53:2, with its roughly cast and badly shaped heads. The enamel decoration makes use of the maximum space available, and the squares and other shapes suggest a derivation from cloisonné. In all six of the bodies illustrated in Fig. 53, decoration is in this manner, and only one, that from Fengate, Peterborough[34] (Fig. 53:8) has reverted to the East Brigantian style, even though the brooch itself is Parisian, judged on head type. The Lamberton Moor brooch[35] (Fig.

53:5) has heads which are out of proportion to the size of the body. In other words, they are too large. Fig. 53:6, from Old Winteringham, Lincolnshire,[36] is not dissimilar in appearance, whilst the enamel colours are unusual, being cream and green. All the brooches of this group rely heavily on coloured enamels for eye appeal. The remaining brooches show a steady deterioration of form and finish. Fig. 53:7, from Kilnsea, Yorkshire,[37] and Fig. 53:9, from Charterhouse, Somerset,[38] on which the number of squares has been increased from four to six, are typical of the devolved form. Curiously, the following two brooches have their heads facing to the left, instead of to the right, which is normal. The Thirst House, Derbyshire[39] brooch, and the Great Bedwyn, Wiltshire[40] brooch are sorry examples of the dragonesque form.

Fig. 51: Dragonesque Brooches, East Brigantian Style 1. Norton, E.R. Yorkshire. 2. Newstead, Roxburghshire. 3. Faversham, Kent. 4. 5. Traprain Law. 6. Margidunum, Nottinghamshire. 7. York. 8. Cirencester, Gloucestershire. 9. Kirkby Thore, Cumbria. (3/4)

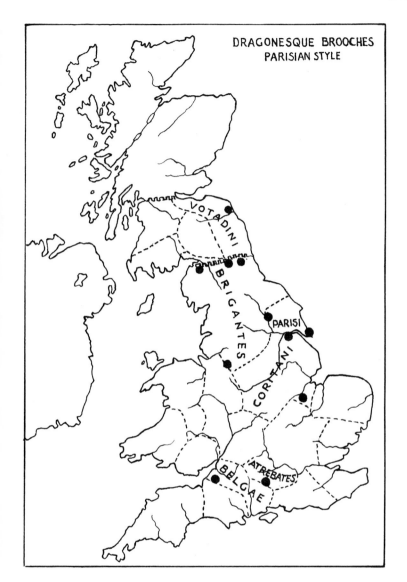

Fig. 52: Dragonesque
Brooches: Distribution
of Parisian Style.

None but the Milking Gap brooch can be dated. It came from a native habitation site, which, amongst other things, yielded glass armlets and pottery with a *terminus a quem* dating of AD 122.[41] The Lamberton Moor brooch was one of several objects in a hoard, amongst which were four bowls of beaten bronze, and two enamelled head-stud brooches.[42] The hoard belongs to the second century, on the evidence of the head-stud brooches and the paterae.[43] On balance it would appear that the Parisian-style dragonesque brooches are of second-century date, making them the latest of their form;

and that the disturbed conditions at the beginning of the century perhaps contributed to their decline.

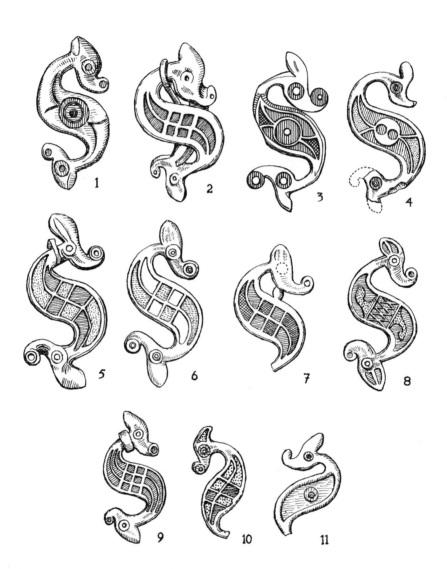

Fig. 53: Dragonesque Brooches, Parisian Style 1. Tanner Row, York. 2. Corbridge, Northumberland. 3. Stanwix, Carlisle. 4. Milking Gap, Northumberland. 5. Lamberton Moor, Berwickshire. 6. Old Winteringham, Lincolnshire. 7. Kilnsea, Yorkshire. 8. Fengate, Peterborough. 9. Charterhouse, Somerset. 10. Thirst House Cave, Buxton, Derbyshire. 11. Great Bedwyn, Wiltshire. (3/4)

(d) Anomalous Brooches

Brooches included under this heading fall into no recognisable category. We begin with the curious instance of four brooches which clearly are the work of the same hand. These are Fig. 54:1, from York; Fig. 54:2, from the Deanery Field, Chester;[44] Fig. 54:3, from Templebrough, Yorkshire;[45] and Fig. 54:4, which was found in the Roman fort at Richborough, Kent.[46] A characteristic common to all is the form of ear-moulding. The upstanding ears must give these brooches some relationship with the East Brigantian style. Body decoration in all cases consists of central panels of lozenge patterns, flanked on both sides by pathetic attempts at scroll patterns.

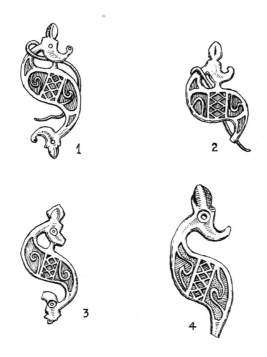

Fig. 54: Dragonesque Brooches, Single Workshop 1. York. 2. Deanery Field, Chester. 3. Templebrough, Yorkshire. 4. Richborough, Kent. (3/4)

All the remaining brooches are illustrated in Fig. 55, and they are something of a mixed bag. One craftsman was clearly a dog lover: for he made the head in the image of a hound. This brooch, Fig. 55:5 was found at Newstead.[47] The rest show a mixture of styles. Fig. 55:1, from the Victoria Cave, Settle,[48] in the very centre of West Brigantian styles, has the body decoration of the Parisian style. The serrated ear is a curious departure from the normal. The largest brooch is Fig. 55:2, from Castlehill Fort, Dalry, Ayrshire,[49] and it is the sole specimen ever found in territory belonging to the Dumnonii. With its central boss, it should be early, yet the heads resemble the Parisian style, and here they have tooth decoration. The shape of the enamel cloisons suggests an attempt at copying lozenges, but not very successfully. Finds at Castlehill fort included fragments of a platter of Sigillata, D. 18–31, and a hilt-guard of Piggott's type IVb, indicating a late-first/early-second-century date. The Corstopitum[50] brooch (Fig. 55:3) came from the 'forum', which also yielded Fig. 50:6. The craftsman's intention here is not clear: very small bosses have been used for eyes; but the swirling body design, though effective and attractive, has no obvious meaning. Fig. 55:4, from Traprain Law[51] is the only plain brooch known. The body shape suggests conjoined cornucopias. Although there is much that is odd about this brooch, the highly stylised design was repeated in part on the Corstopitum[52] belt-mount (Fig. 15:10). One half of a brooch from Traprain Law (Fig. 55:6), from the lowest level, shows an omission of the neck. The Corinium, Gloucestershire[53] brooch (Fig. 55:7) is a neat specimen, supposedly in the Parisian style, but the head-shape is wrong. Fig. 55:8, from South Shields,[54] is not really a dragonesque brooch, because it is more Teutonic in feeling, and is unlikely to have been in the dragonesque style. The Braughing, Hertfordshire[55]

Fig. 55: Dragonesque Brooches, Miscellaneous 1. Victoria Cave, Settle. 2. Castlehill Fort, Dalry, Ayr. 3. Corstopitum, Northumberland. 4. 6. Traprain Law. 5. Newstead, Roxburghshire. 7. Cirencester, Gloucestershire. 8. South Shields, Co. Durham. 9. Braughing, Hertfordshire. 10. Rudston, Yorkshire. (3/4)

brooch, fashioned out of a flat piece of bronze, belongs to no recognisable form or style, though meant to be a copy of the dragonesque form. Finally, at Rudston, Yorkshire[56] somebody wrote the obituary of the dragonesque brooch form in a length of bronze wire.

NOTES

1. Collingwood, 'Romano-Celtic Art in Northumbria', p. 53.

2. *Yorks. Arch. Journ.*, XXXIV (1939), p. 137, Fig. ii:26.

3. *Proc. Univ. Durham. Phil. Soc.*, IX, p. 192, Fig. 1:7.

4. Munro, *Lake Dwellings of Europe* (1890), p. 424, Fig. 153. An associated find was a piece of Samian bowl of Form D. 37, of second-century date.

5. Curle, *A Roman Frontier Post and its People*, pl. LXXXV:6.

6. One was found in the Victoria Cave, Settle. Collingwood[1], p. 53; *V.C.H. Derbyshire*, I, p. 239.

7. Collingwood completely misunderstood the form. He described it as a clumsy thing with meaningless mouldings at head and foot.

8. Clarke, 'The Iron Age in Norfolk and Suffolk', p. 55, Fig. 10:1.

9. Ibid., p. 55.

10. *Arch. Journ.*, XXX (1873), p. 59.

11. In addition to the sources mentioned, there were copper deposits just north of the Tweed, near Cockburn's Law, in the Lammermuirs. A circular ingot of almost pure copper, about 36 lbs weight, was found at Carleton, Wigtownshire.

12. A. Raistrick, 'Iron Age Settlement in West Yorkshire', *Yorks. Arch. Journ.*, XXXIV (1939), p. 137, Fig. ii:19.

13. R.G. Collingwood, *Archaeology of Roman Britain*, Fig. 64:112.

14. W. Boyd Dawkins, 'Results . . . out of the Victoria Cave in 1870', *Journ. Anthrop. Inst.*, I (1871–2), p. 62, pl. facing p. 1, no. 3.

15. Collingwood[1], p. 57.

16. F. Haverfield, 'Small Objects, 1910 Excavations at Corstopitum', *Arch. Ael.*[3], VII (1911), p. 188, Fig. 28.

17. R.E.M. Wheeler, 'Segontium and the Roman Occupation of Wales', *Y Cymmrodor*, XXXIII (1923), Fig. 56:3.

18. Curle, 'Objects of Roman and Provincial Roman Origin found in Scotland', p. 343, Fig. 37.

19. *Arch. Ael.*[3], V (1909), p. 422.

20. R.A. Smith in *Proc. Soc. Antiq.*[2], XXII (1907–8), p. 59, Fig. 4.

21. *Proc. Arch. Inst. of Gt Britain* (1846), II, pp. 17, 35, pl. 1:4. See also *BM Guide to Antiquities of Roman Britain*, Fig. 12:47.

22. Curle[5], p. 319, pl. LXXXV:7. Also *Journ. Roman Studies*, III, p. 105, Fig. 12.

23. *BM Guide*[21], Fig. 11:25.

24. A brooch, identical with the Colchester specimen, was found at Traprain Law; *Proc. Soc. Antiq. Scot.*, LVIII (1923–3), p. 251, Fig. 9:1. See also Fig. 18:7,

this book.

25. J.P. Bushe-Fox, 'Third Report on the Excavations on the Site of the Roman Town at Wroxeter, Shropshire', *Soc. Antiq. Research Report*, IV (1914), p. 24, pl. XVI:9.

26. Unpublished.

27. E.g., Curle[18], p. 332. Absence of a common form at Traprain Law makes local manufacture unlikely.

28. Smith[20], Fig. 2a, b and c.

29. *Antiq. Journ.*, XXXI (1951), p. 40, Fig. 2:28.

30. *Arch. Ael.*[4], XV (1938), p. 342. *Trans. C. W.* (new series), XIX.

31. Kilbride-Jones, 'The Excavations of a Native Settlement at Milking Gap, High Shield, Northumberland', p. 342.

32. Jacobsthal, *Early Celtic Art*, p. 78.

33. Forster and Knowles, 'Corstopitum: Report on the 1914 Excavations', p. 245, no. 1.

34. F.W. Feachem, 'Dragonesque Fibulae', *Antiq. Journ.*, XLVIII (1968), p. 100, Fig. 2:46.

35. *Proc. Soc. Antiq. Scot.*, XXXIX (1904–5), p. 375, Fig. 5. Also Curle[18], Fig. 48:2. It is incorrectly drawn in *Proc. Soc. Antiq.*[2], XXII (1907–8), Fig. 59:5.

36. Feachem[34], Fig. 2:48.

37. *Roman Malton and District Report*, no. 5 (1935), p. 95.

38. *V.C.H. Somerset*, p. 337, Fig. 92.

39. J. Ward, 'Recent Cave Hunting in Derbyshire', *The Reliquary* (New Series), III (1897), p. 94, Fig. 10.

40. *Wilts. Arch. Mag.*, Liii, p. 376.

41. Kilbride-Jones[31], p. 342.

42. Curle[18], p. 332.

43. *Proc. Soc. Antiq. Scot.*, LXII (1927–8), p. 246.

44. *Annals of Archaeology and Anthropology*, XV, p. 17, pl. VII:3.

45. T. May, *The Roman Fort at Templebrough* (1922), p. 71, pl. XIV:1.

46. Bushe-Fox, 'Third Report on Excavations of the Roman Fort at Richborough, Kent', p. 77, pl. IX:12.

47. Curle[18], p. 398, Fig. 1.

48. Boyd Dawkins[14], pl. facing p. 1, no. 7.

49. J. Smith, 'Excavation of the Forts at Castlehill, Dalry, Ayrshire', *Proc. Soc. Antiq. Scot.*, LIII (1918–19), p. 128, Fig. 4:1. Also Curle[18], p. 378, Fig. 59:1.

50. Haverfield[16], Fig. 20.

51. *Proc. Soc. Antiq. Scot.*, LVI (1921–2), p. 251, Fig. 28:4.

52. Forster and Knowles[33], p. 245, pl. I:III.

53. J. Buckman and C.H. Newmarch, *Illustrations of the Remains of Roman Art in Cirencester* (1850), pl. X:7, but incorrectly drawn.

54. Smith[20], p. 62, Fig. 6; *Arch. Ael.*[3], IV, p. 357.

55. Collingwood[1], Fig. 11:a.

56. Feachem[34], Fig. 2:51.

20. Horse Trappings

The effects of Romanisation will now have been made clear: a narrowing of production to the manufacture of trinkets and souvenirs; a rapid deterioration in the art; a general malaise prompted by commercial considerations. The position has still to deteriorate further. But it is interesting to note that this situation had been arrived at in rather less than 100 years; and for Brigantia, at any rate, the process started later, but acceleration was faster once we are into the second century.

North of the Brigantian border, conditions were not quite so bad — at least for the time being. Here there is an observed persistence in the manufacture of enamelled trappings into the second century. Numbers are small, but the quality of workmanship remains high. But even here the age-old motifs were being dropped in preference for a new style of decoration, some of which we have already seen on some dress-fasteners. Thus we have a suggestion of contemporaneity, but a certain amount of caution is called for, seeing that we are here dealing with fringe areas.

There are terrets and bridle-bits, and their Celtic character is maintained. The terrets are elliptical, and of equal cross-sectional area throughout, meaning that there is no expansion: but at the bottom there is the attachment bar flanked by mouldings in the usual manner (Fig. 26:5). Attached to the loop are 'platforms', which are vehicles for enamel decoration. It will be remembered that previously we have considered knobbed terrets and lipped terrets, and now we have platform terrets. The platforms are square. A pair of platform terrets (Fig. 26:5) was found at Birrens, Dumfriesshire,[1] and they are a good starting point from which this discussion can be pursued, since they were stratified. They came from a burnt deposit, on level II on site VIII, a homogeneous deposit that is assignable to the end of the period which closed shortly before AD 158. These terrets are therefore late, and the date, in the middle of the second century, gives some measure of the persistence of equitation amongst the native population up till this time. 158 was also the year of the Brigantian revolt, and the presence at Birrens of these terrets must indicate the use of chariots. These terrets, together with the cheek-piece (Fig. 15:5) may be relics of Brigantian casualties in face of the Romans. So far as the reconstruction of Birrens in 158 is concerned, this event must be related to the history of the immediate neighbourhood, as Birley has remarked.

The design of the Birrens terrets must represent the ultimate. The enamelled decoration of the square platforms compares well, both in style and in format, with that on some dress-fasteners, notably the one from York (Fig. 45:10). The same pattern appears on a square looped stud from Kirkby Thore, Cumbria,[2] and again on the Birrenswark, Dumfriesshire bridle-bit (Fig. 56:2), in slightly modified form, and also the elongated strap junction from High Rochester, Northumberland.[3] Another strap junction, this time from Inchtuthil, Perthshire,[4] has a similar pattern, but without the squares; and in this respect this absence could indicate that this is an earlier version, since Inchtuthil was a legionary fort, established in 86 and abandoned *c*. AD 90. Therefore, it would seem that the pattern in question was current in the first half of the second century, a date which seems appropriate to the square dress-fastener as well.

Certain terrets possess a regional character. Simple looped terrets were common over most of Britain, except for Wales, occurring from Cornwall in the south-west to the Lowlands in the north, with a stray found at Ardoch. But ribbed terrets and crescent terrets were confined to south of the Jurassic Ridge. Lipped terrets were similarly common south of the Jurassic Ridge, but they were found in numbers only at Stanwick, Yorkshire, and there was a stray at Newstead. Knobbed terrets clearly belong to John Gillam's inter-isthmus province, with an overspill into Brigantia; but platform terrets look like being north Brigantian, though there is a certain emphasis on manufacture north of the Wall. There is a link between High Rochester, Newstead and Traprain Law, and a stray was found at Balmuildy, Lanarkshire. But these Scottish-based terrets are earlier, for at Traprain Law they are of first/early-second-century date, but the exact attribution remains uncertain. But the largest and the best decorated platform terrets are in Brigantia; and the Birrens terrets remain the ultimate.

Yet the form we are talking about seems unlikely to have originated in Brigantia. A clue to the source of inspiration is provided at Balmuildy, Lanarkshire[5] in the terret found there. There is not very much of it left, except for one platform and part of the hoop; but the similarity between this and the terrets, particularly in the matter of decoration, from Sahan Toney, Norfolk,[6] cannot be denied. The Sahan Toney decoration consists mostly of lozenges, but squares are also present. The Balmuildy terret was found in a post-hole in the barracks block XI, and the pottery here is mostly Flavian, though the building itself was Antonine. Sahan Toney is in the territory of the Iceni, and the hoard was probably buried, like most of the others in East Anglia, about AD 60, after the failure of the Boudiccan revolt. The time difference involved is not very great, and the patterns may have reached Brigantia on someone fleeing from the Romans. However, there is sufficient evidence to show whence the cellular form of enamel decoration was obtained, and once it reached the north it was liberally applied to a variety of articles, dress-fasteners among them; and the original layout was improved upon by the Brigantes through the addition of interlocked triangles. The whole history of this form of decoration shows a parallel eclipse of the underlying motifs at one time common in Celtic art. Either this was a gesture to the Romans, or else the new style was adopted because it called for less skill on the part of the metalworkers.

However, in the north the art was not yet moribund. There is still vigour shown in the decoration on the bridle-bit from Birrenswark, Dumfriesshire,[7] (Fig. 56:2), though the old art forms are about to be swamped by the cellular patterns just discussed. On this bridle-bit the new style is represented in all its cellular glory of coloured enamels, striking to the eye, perhaps, and in that respect its main appeal may lie. Here there are squares and interlocked triangles, and incorporated with these is the last remnant of Celtic art, made up of twin petals or seed-shaped cloisons, which are related to swash Ns, to be seen again in Fig. 56:6. But, on this Birrenswark bridle-bit the related swash N pattern is given pride of place, as a final protest, perhaps; but it is surrounded on three sides by cellular patterns intent on the final take-over. This survival of the old swash N pattern is a little strange, for it is not a localised affair. Precisely in the Birrenswark form, it occurs twice

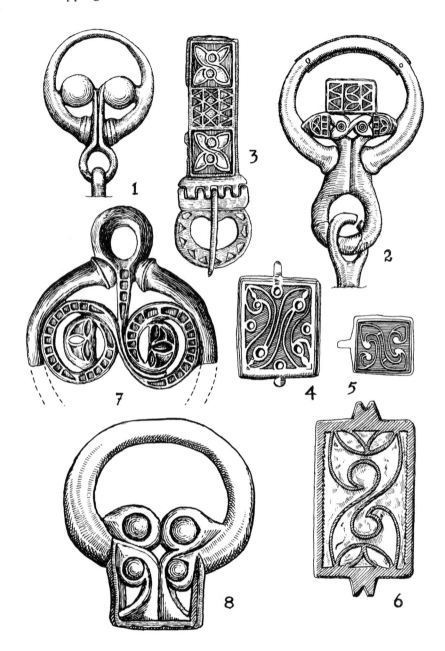

Fig. 56: 1. Bronze Bridle-bit, Middlebie, Dumfries. 2. Enamelled Bronze Bridle-bit, Birrenswark, Dumfries. 3. Enamelled Bronze Belt-end, Caerleon, Monmouth. 4. Enamelled Bronze Seal-box Lid, Lincoln. 5. Blue Enamelled Seal-box Lid, Cilurnum, Northumberland. 6. Enamelled Bronze Lid, Corstopitum, Northumberland. 7. Fragmentary Enamelled Bronze Bridle-bit, Seven Sisters, Neath, Glamorgan. 8. Bronze Mount, Stanhope, Peebles. (3/4)

on a belt-end found in the legionary fortress at Caerleon, Monmouth.[8] This belt-end is of a Roman military type, commonly worn in the first century AD.[9] However, this statement does not imply that the origin of the piece was Roman: its composition (Fig. 56:3) clearly shows its decoration to have been native designed, but the overall composition also demonstrates the native attitude to a changing market. Between the two swash Ns represented here there is a panel with central lozenges flanked by interlocked triangles, in two colours of enamel. Although this belt-end was found in disturbed soil in Barrack V, Lady Fox is of the opinion that the available evidence suggests a date for this piece within the period AD 50–130. An almost identical belt-end was found at Richborough, Kent,[10] but it is unstratified.

Even with the few examples extant, it is clear that swash Ns must have been more widespread than previously thought possible, with examples cited in this chapter having come from such widely spaced areas. Therefore, they must represent a period style. Yet, swash Ns are essentially at home in the north, where they made their initial bow at Plunton Castle (Fig. 15:2). There are enamelled seal-box lids bearing this pattern; of such is the one from Lincoln[11] (Fig. 56:4), and there is another like it in the Mersey county Museum. Other examples occur on the seal-box lid from Cilurnum (Fig. 56:5), on which the enamel background is in blue. Perhaps the ultimate in a devolved form appears on the cover from Corstopitum[12] (Fig. 56:6). And with this we come back to the pattern on the Birrenswark bridle-bit. The swash N is thought by Lady Fox to be a variation on the broken-back scroll,[13] but this is unlikely. But it must be remembered that broken-back scrolls were known in Galloway, at Lochar Moss, and there was something approaching them on the Stichill collar. However, Leeds' view is to be preferred: he saw the pattern as a pseudo-calligrapher's florid conception of the letter N.

Birrens has provided a *terminus ad quem* date of AD 158 for the type of cellular decoration appearing on terrets, bridle-bits and dress-fasteners. But this was the year of the Brigantian revolt, which Collingwood[14] blamed (incorrectly) for the cessation of production of cheap trumpet brooches, which were the metalworkers' staple wares. But there is a noticeable fall-off of enamelling at this time, since the only objects that can be put into a later period than this are a few knobbed terrets, and they are entirely plain: for example, Poltross Burn, Cumbria,[15] lost some time between 200 and 320; or Traprain Law,[16] supposedly found in a level dated by late-second-century coins. A fragment of a plain platform terret from the same oppidum is said to be of the same period. However, one must take the precaution of stating that these attributions are slightly suspect, though the fact remains that, after 158, there are no outstanding enamelled items. With the gradual erosion of privileges, due to repeated skirmishes with the enemy, the native craftsmen must have found it increasingly difficult to practice their craft. The Fendoch sword[17] makes it clear that some craftsmen had gone to work for the Romans, and that this had occurred as early as AD 82–6. Others poured out disc-brooches and fake trumpet brooches as a profitable method of earning a living. It is a little surprising that the art lingered on for as long as it did, finally to receive the *coup de grace* in 196, when the northern tribes invaded the province and destroyed the workshops.

NOTES

1. Eric Birley *et al.*, 'Excavations at Birrens, 1936-37', p. 337, Fig. 38:1 and 2.

2. *Proc. Arch. Inst.*, York Volume (1846), pl. II, pl. 1.

3. Henry, 'Emailleurs d'Occident', p. 105, Fig. 21:6.

4. Macgregor, *Early Celtic Art in North Britain*, no. 32.

5. S.N. Miller, *The Roman Fort at Balmuildy* (Maclehose-Jackson, 1922), pl. LI:4, and LIII:2.

6. *V.C.H. Norfolk* (1900), p. 273.

7. Anderson, *Scotland in Pagan Times*, p. 123, Fig. 101.

8. Fox, 'The Legionary Fortress at Caerleon, Monmouth', p. 128, Fig. 6:10.

9. Bushe-Fox, 'Excavations . . . at Richborough, Kent' (Fourth Report), p. 123.

10. Ibid., pl. XXXIII:73.

11. *BM Guide to Antiquities of Roman Britain*, p. 78, Fig. 40:10.

12. Note by Haverfield in Forster and Knowles, 'Corstopitum: Report on the Excavations in 1910', p. 178, 188, Fig. 30.

13. Fox[8], p. 128.

14. Collingwood, 'Romano-Celtic Art in Northumbria', p. 57.

15. J.P. Gibson and F.G. Simpson, 'The Milecastle on the Wall of Hadrian at Poltross Burn', *Cumb. & West. Arch. Soc.* (new series), XI (1911), p. 442, Fig. 21:13.

16. E. Fowler (nee Burley), 'A Catalogue and Survey of the Metalwork from Traprain Law', *Proc. Soc. Antiq. Scot.*, LXXXIX (1955–6), p. 194, no. 350.

17. Richmond and MacIntyre, 'The Agricolan Fort at Fendoch', p. 146, pl. LX:1.

21. Pins, and the Atlantic Province

Childe's basic tenet was that objects, whatever their form, are the material expression of cultures.[1] On grounds of that statement, and because pins remain the chief material expression (in metal) of the Atlantic Province culture, then technology must have been at a very low ebb in that area. The Atlantic Province is defined as that region of Scotland in the west, stretching from the Mull of Kintyre to John O' Groats, and including all the western isles and the Orkneys. Except for the Orkneys, which, as we know from our Tacitus, were added to the Roman Empire, the Province was remote from military areas, and therefore from Romanisation; and for that reason maintained a native culture which, even if primitive by the standards of the times, was nonetheless indigenous. Pins provided a link between this and the cultures of eastern Scotland, and this appears to have resulted from acquisitions made by the broch people of the north.

In this Atlantic Province there are numerous finds of Roman objects, including Sigillata, which bespeak of other connections, perhaps with the south-west Scottish cultural areas; so that there must have been a lot of movement up and down the western seaboard. Roman pottery sherds are particularly noticeable in Argyll, well placed for the quick kill; but also they occur in Tiree, Skye, North Uist and Harris. There are concentrations south of John O' Groats and in the Orkneys. The general picture is one of contact with the enemy, or with the country's Romanised citizens; but whether culturally or forcefully is not evident. It is more than passing strange that no object, which can be termed Celtic because of its decoration, has so far been found in the Atlantic Province. Instead, the metalwork is made up chiefly of spiral rings and pins, neither of which sounds very exciting. Both are Traprain Law products, the largest number having been found there; and although the distribution[2] of the former makes it clear that some contact with eastern Scotland must have been maintained, for example, with East Lothian, Roxburgh, Fife and Angus, via the Lowlands, and Ayrshire with Argyll, not a single work of art has come to light. The story concerning projecting ring-headed pins is similar, though the distribution remains coastal.

In this context, therefore, pins and their development become a matter of importance, in that, to a major extent, they may be regarded as a native development. And pins, perhaps more than any other object, have a long history, some forms seeing a slow but not always continuous development from late Bronze Age times, through the whole Occupation period up till and including the Dark Ages, and perhaps beyond. If the end product of so long a development is simple and, to some, disappointing, this is because a pin does a simple job, only the head being visible; and only for special wear on special occasions is it likely that ornamental heads would be deemed necessary. Even the inexperienced metalworker would be able to make the simple forms of pin, such as the projecting ring-headed type.

189

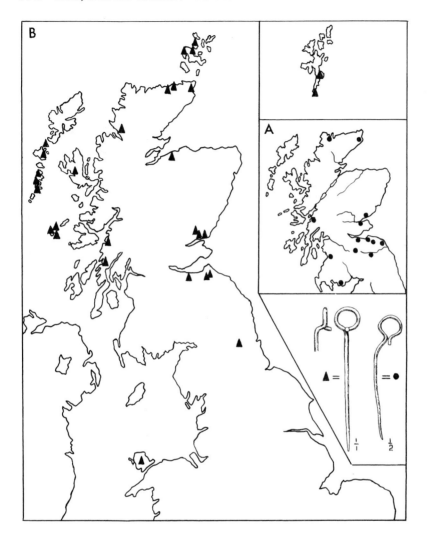

Fig. 57: Maps showing Distribution of (A) Ring-headed Pins, and (B) Projecting Ring-headed Pins in North Britain.

Even with such simple objects as pins, arguments arise as to what suggested this form in the first place. Look at Figs. 57–60 – some would advocate derivation from the sunflower pin;[3] but, why should basically simple forms have to be derived from anything? If we have to commit ourselves, then we would say that the ring-headed pin is ancestral to the simple projecting ring-headed pins (Fig. 57 and Map A). The distribution of one is seen to be similar to the distribution of the other (compare Map B with Map A). These similarities of distribution form the basis of our argument; in both cases the largest numbers occur on the south banks of the Forth, with a concerted involvement at Traprain Law. Ring-headed pins, numerous in the south, underwent literally no development over a period of about half a millennium. By any account that is a long time, and absence of development renders the pin useless for dating purposes. Wheeler[4] regarded them as a British invention, but based upon the Hallstatt swan-necked pin. Ancestor hunting gives the metalworkers credit for having no brains. The involvement of Traprain Law in the manufacture of pins, not only of the projecting ring-headed type, but of other forms as well, is well authenticated by the discovery there of clay moulds for casting them. The Votadini could have been responsible for the development of this form out of the ring-headed variety, which in bronze at Maiden Castle was dated to the first century AD. In view of their numbers and their distribution, there is no reason why the ring-headed pin should not have given way to the projecting form, as a matter of common sense and natural development; for the simple reason that it is better to have the head lying close to and flat against the cloth, as is the case with the Irish *fainne*.

Perhaps development is not so important as the fact that the form is there, scattered around Scotland in a coastal distribution, as seen on the Map B, Fig. 57. Of these, the most southerly occurrence is a single pin from Anglesey. The situation of this pin, so far from home, only adds to the mystery as to why this form has not yet been found in Ireland. The most primitive-looking of these projecting ring-headed pins is that from Dun Fheurain, Gallanach, Oban, Argyll, at which site there was a refuse-heap at the base of the rock. This refuse-heap has yielded pins, needles, ring-headed pins of bronze and of iron, a sword blade, coarse pottery and a small fragment of Sigillata — a wide variety of objects but none capable of setting an archaeologist's imagination afire. Another association of pins with Sigillata was at the Midhowe broch, Rousay, Orkney.[5] There were three pins here, and two are shown in Fig. 58:5 and 7.

The association of these pins with Sigillata is interesting, since it has been noted that spiral rings and pins of the variety under consideration often come from areas where Roman objects are found. And strangely, these are all coastal sites, or they are within easy reach of the sea; and Stevenson[6] was right to draw attention to an association with the broch culture. The association of pins with the broch culture would explain the absence from Caledonia of most forms of pin discussed here. An association between the broch people and the Votadini of Traprain Law must also be mooted, in view of the occurrence of these pins and of zoomorphic brooches, both Traprain Law products, in the northern brochs. But if Caledonia was isolated, it nevertheless had its own workshops; and it may have been the Caledonians who overran Traprain Law in 196, in retaliation for one reason or another. Therefore, the pins must belong to the second century, and this date is confirmed by their third-level associations at Traprain Law. Although there are specifically four occupation levels at Traprain Law, the two lowest (levels 4 and 3) show a happy balance of Roman, native and Romano-British, whereas the two upper levels (levels 2 and 1) are essentially 'native', as Elizabeth Fowler has pointed out.[7] Between these two groupings there is a complete break in cultural continuity, which resulted from the overrunning of Traprain Law by the northern tribes in AD 196. After this disaster, the oppidum remained unoccupied for some 30 or 40 years. This division of the occupation into two periods, rather than four levels, is a more satisfactory background to the continuing story of the pins.

For it is the Traprain Law story which provides the only credible background to pin development. A glance at all the pins of Fig. 58 is sufficient to show that the only noticeable development is that the pins, from having been made from circular wire, in one or two cases have bevelled heads (Fig. 58:7). At Ness broch the bevelled head appears to be cast, but the pins from the Culbin Sands and Dunadd (Fig. 58:10 and 11) have heads bent to shape. The Dunadd pin might be thought of as being late, in view of the supposed association of this site with the capital of Dalriada; but it must be remembered that during the 1929 excavations[8] some sherds of Sigillata were found here, proving earlier occupation.

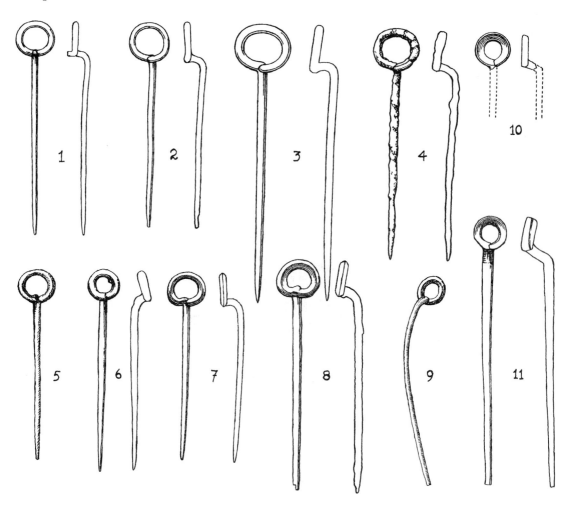

Fig. 58: Bronze Projecting Ring-headed Pins 1. Dun Fheurain, Gallanach, Oban, Argyll. 2. 10. Culbin Sands, Morayshire. 3. Anglesey. 4. The Laws, Monifieth. 5. 7. Midhowe, Rousay, Orkney. 6. 9. Traprain Law. 8. Ness Broch, Caithness. 11. Dunadd, Argyll. (3/4)

The Traprain Law story is progressive. It is progressive in the sense that new forms appear; but there are some like the ribbed elliptical form from Aikerness broch, Orkney (Fig. 59:1) and its nearest analogous example from Keady, Co. Armagh (Fig. 59:2) which are not represented at the oppidum. New pin styles appear; these are the rosette type, with beads all round the ring (Fig. 59:3–7) and the half-rosette type, in which the beads occur only on the upper half, the lower half being in the form of a semicircular plate (Fig. 60:2–5). Both types were made in the second century, as evidence for which there are the moulds from the first period, or third level. These are native productions, uninfluenced by outside cultural trends, and quite unique in their way, being representative of purely British metalwork at this time. A rosette-type pin was found in level five[9] (Fig. 59:3), but this kind of association is exceptional, and the pin is out of normal context, since Fig. 59:4[10] and the mould, Fig. 59:9, both came from the third level, which is more in keeping with what one would expect. The rosette type consists always of six beads without separating fillets. There is never any variation on this number, not even when manufacture of the type was resumed in Traprain Law's second 'native' period in the third century, following upon the reoccupation of the oppidum. Fig. 59:5 and 6 both came from the second level, and were thus manufactured during the purely native phase of occupation. Perhaps this continuity of form, after a gap of some 30 or 40 years, indicates a return of some of the old metalworkers. There are eight rosette pins from Traprain Law, whilst others come from the Sculptor's Cave, Covesea[11] (Fig. 59:7) and Aesica[12] (Fig. 59:10). Unfortunately, the Aesica pin was not stratified; but Aesica was not built before AD 128, since it was a secondary Wall fort, and maintained on a care-and-maintenance basis. However, it is known that the granary was restored by Maximus in 225, because it was said to be old and ruined — perhaps as a result of some frontier disaster — and later the fort was destroyed by the Picts *c.* 306.[13] A curious variation on this rosette type is Fig. 59:8, also from Traprain Law. There is another like it from Tentsmuir, Fife (Fig. 59:11).

The history of the half-rosette type of pin is very similar. As a type, it too was manufactured in the second century, as the mould, Fig. 60:2, testifies. This mould was found in the third level at Traprain Law,[14] and it is intended for a pinhead with six beads at the top. After the reoccupation of the oppidum in the third century, as was the case with the rosette type, manufacture was resumed, but the number of beads was reduced to five (Fig. 60:3, from the second level); then to four beads (Fig. 60:4 and 5); and finally to three beads (Fig. 60:2a). The last drawing is from a cast taken from a mould which was found in the top level at Traprain Law. Fig. 60:4 is a four-beaded pin recovered from the Sculptor's Cave, Covesea,[15] whilst the other four-beaded specimen is from Corstopitum,[16] unfortunately unstratified. This gradual but persistent devolution is highly important, since, once the number of beads had been reduced to three, it led to the emergence of the three-fingered handpin as a definite form. A fully evolved handpin, complete with three fingers instead of three beads, came from the second level at Traprain Law[17] (Fig. 60:9). Here is surely the answer to the oft-asked question: where did the handpin originate? In this respect, the term 'proto-handpin' is used too indiscriminately. Stevenson proposed this term for Fig. 60:9, but when examined this pin will be found to be as mature as are most of the Irish three-fingered handpins; and in this respect we believe that Traprain Law was first with the idea.

Actually, the term 'proto-handpin' would be better applied to something like Fig. 60:8, which is a half-rosette pin in silver from Ireland;[18] but it has been given the Irish treatment, in that the beads are no longer round, but elongated, which resulted from the inclusion of fillets. This development was far reaching, for it led to the proto-handpin development of Fig. 68. It was a far travelled type; apart from Covesea and Corstopitum, one was found on Berneray, North Uist.[19] Irish workmanship was always superior to anything comparable at Traprain Law, with a greater ability to adapt and to modify; but the Votadini were more original in the matter of designing prototypes.

There is now a broader and less regional distribution. For instance, the pins figured in

Fig. 61 have an Atlantic distribution; Fig. 61:1 comes from a wheelhouse at Bruthach a Tuath, Balivanich, Benbecula,[20] a pin which Stevenson[21] looked upon as the ultimate in degeneration of the ibex-head. There were, however, ring-headed pins in Ireland having three pellets or beads,[22] as is the case here. One with four beads on the ends of separate stalks (Fig. 61:2) was found on Balavullin Sands, Tiree,[23] and a plate form, with a ring on the back for a chain, was found in Waterford, Ireland (Fig. 61:3). Though the style of these pins is individual, and it represents a definite trend in pin-making, their period is a matter for conjecture. But a remote relationship between Ireland and the Atlantic Province is beginning to emerge. This has been noted before. The heads on the swivel-ring from a'Chrois, Tiree and Vallay, North Uist, are definitely similar to those on the Irish mirror handle (Fig. 33:4 and 3).

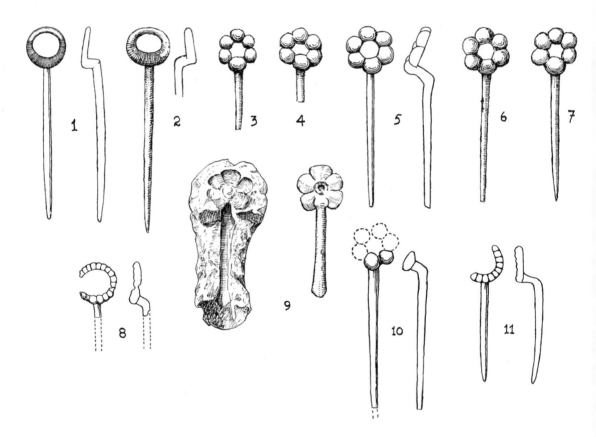

Fig. 59: *Bronze Ribbed Elliptical Pins* 1. Aikerness Broch, Orkney. 2. Keady, Co. Armagh. *Rosette Type Pins* 3.–6. Traprain Law. 7. Sculptor's Cave, Covesea. 9. Clay Mould for Casting Rosette-type Pin, Traprain Law. 10. Aesica, Northumberland. *Beaded Type Pins* 8. Traprain Law. 11. Tentsmuir, Fife. (3/4)

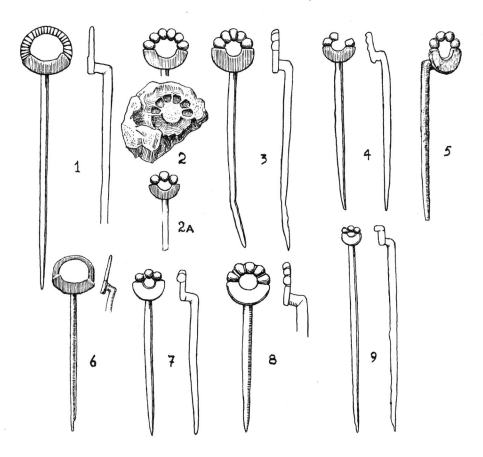

Fig. 60: 1. Half-beaded Pin, Bowermadden Broch, Caithness. 6. Similar, but Plain, Lydney, Gloucestershire. *Bronze Half-rosette Type Pins* 2. Clay Mould for Casting Pin, Traprain Law. 3. Traprain Law. 4. Sculptor's Cave, Covesea. 5. Corbridge, Northumberland. 7. No Locality (Mus. No. GT 237). 8. Silver, Ireland. 9. Traprain Law. *Bronze Three Fingered Handpin* 10. Traprain Law. (3/4)

Another type of pin was common at the Sculptor's Cave, Covesea, but it is unrepresented at Traprain Law. The constituents of this new series have circular heads, ribbed at the top, but having beads, usually three in number, on the lower part of the ring. Their devolution led to the emergence of what has been called the ibex-headed pin, which Reginald Smith once dated to the first century BC.[24] The first pin illustrated here (Fig. 61:4) is from North Berwick,[25] whilst Fig. 61:5, 6, 7, 9 and 11 are from the Sculptor's Cave, Covesea. Whereas the North Berwick pin has fillets between beads, the rest have not, except for no. 11. All the Covesea pins are essentially of the same small form, but with slight differences in bead representation. In the case of Fig. 61:5, the beads are round, but on the smaller pins they tend to become almost wedge-shaped on the inside, and perhaps this is how no. 7 evolved, the one pin in this collection which everyone regards as being ibex-headed. However, there is another aspect of Covesea development which should not be passed over lightly, and this is the sudden and noticeable increase in the size of the centre bead in the case of Fig. 61:6. This single bead is prominent here and rounded, and out of proportion with its neighbours. In the matter of style, this pinhead must be compared with Fig. 61:13, which is from Ireland. At the base of this Irish pinhead there is a round bead, comparable with Fig. 61:6; but in the case of the Irish specimen this bead is divorced from its neighbours. On the other Irish pin, Fig. 61:12, the tops of the beads have been flattened. Both Irish specimens are accepted ibex-headed pins, so that the origin of the ibex-head should now be clear. The classic and oft-illustrated example of an ibex-headed pin is the one from Sandy, Bedfordshire[26] (Fig. 61:8). Others have been found in North Uist[27] (Fig. 61:14) and at Dunfanaghy, Co. Donegal.[28] Complementary with the Covesea series is a pin (Fig. 61:10) found at Lydney, Gloucestershire,[29] but unfortunately not dated, though it is likely to be of fourth-century date here. (Further discussion of the date of ibex-headed pins will be found on p. 248.)

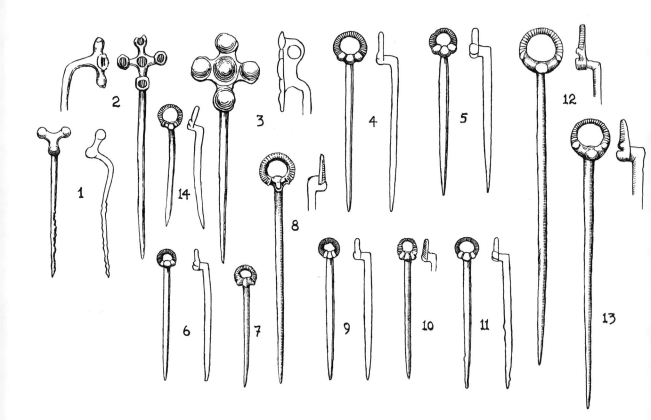

Fig. 61: *Miscellaneous Bronze Pins* 1. Bruthach a Tuath, Balivanich, Benbecula. 2. Balavull-in-Sands, Tiree. 3. Co. Waterford. *Ribbed and Beaded Type Pins* 4. North Berwick. 5.—7. 9. 11. Sculptor's Cave, Covesea. 10. Lydney, Gloucestershire. *Ibex-headed Pins* 8. Sandy, Bedfordshire. 12. 13. Ireland. (3/4)

This short exercise has demonstrated that pins, so often despised by archaeologists, also have their story to tell. It is not a very specific story, and this is why the coastal distribution of the projecting ring-headed pins remains a bit puzzling; yet demonstrates the existence of a pin-conscious area in the Atlantic Province. Traprain Law emerges as the manufacturing centre for most of these pins in their various forms, though whether or not the Atlantic Province pins were imported from here must remain a moot point. Pins are so simple they could be made almost anywhere. But form remains the dominant factor, and in this respect it brings together almost all the areas mentioned, from the east of Scotland to the Atlantic Province, and to some extent Ireland: and Columba's going to Iona in 563 was made easier by the common bond that must have existed between the peoples of the areas mentioned.

NOTES

1. V. Gordon Childe, *Piecing Together the Past* (Routledge, 1956), p. 16.

2. D.V. Clarke, 'Small Finds in the Atlantic Province', *Scottish Archaeological Forum*, 3 (1971), p. 22.

3. Summarised in Clarke[2].

4. Wheeler, 'Maiden Castle', p. 267.

5. J. Graham Callander and W.G. Grant, 'The Broch of Midhowe, Orkney', *Proc. Soc. Antiq. Scot.*, LXVIII (1933–4), p. 500, Fig. 44:1–3.

6. R.B.K. Stevenson, 'Pins and the Chronology of Brochs', *Proc. Preh. Soc.*, XXI (1955), p. 282.

7. Fowler (nee Burley), 'A Catalogue and Survey of the Metalwork from Traprain Law', p. 131.

8. *Proc. Soc. Antiq. Scot.*, LXIV (1929–30), p. 124.

9. *Proc. Soc. Antiq. Scot.*, LVI (1921–2), p. 221, Fig. 20:1.

10. Ibid., p. 236, Fig. 29:4.

11. S. Benton, 'Excavation of the Sculptor's Cave, Covesea, Morayshire', *Proc. Soc. Antiq. Scot.*, LXV (1930–1), p. 194, Fig. 16.

12. Newcastle Museum, 1894–5, or 1897.

13. *Arch. Ael*[2], XXIV (1903), pp. 31, 51.

14. *Proc. Soc. Antiq. Scot.*, LIV (1919–20), p. 80, Fig. 14:2.

15. Benton[11], p. 194.

16. Haverfield, 'Small Objects, 1910 Excavations at Corstopitum', p. 188, Fig. 34.

17. Curle and Cree[14], p. 88, Fig. 19:4.

18. National Museum of Ireland, no. 1920:53.

19. R.E.M. and T.V. Wheeler, 'Report on the Excavation of the Prehistoric, Roman, and post-Roman site at Lydney Park, Glos.', *Soc. Antiq. Research Comm. Report*, IX, p. 83, Fig. 18.

20. *Proc. Preh. Soc.*, XVIII (1952), p. 184.

21. Stevenson[6], p. 291.

22. *BM Guide to Early Iron Age Antiquities*, p. 97, Fig. 106.

23. L.M. Mann Collection, Glasgow.

24. R.A. Smith, *Proc. Soc. Antiq*[2]., (1903–5), p. 350.

25. *BM Guide*[22], p. 97, Fig. 111.

26. Smith[24], p. 350.

27. J. Close Brooks and S. Maxwell, 'The Mackenzie Collection', *Proc. Soc. Antiq. Scot.*, 105 (1972–4), p. 287, Fig. 1:968.

28. *Ulster Journ. Arch*[3], 13 (1950), pp. 54–6.

29. Wheeler[19], Fig. 18.

22. Zoomorphic Penannular Brooches

The second century, as we have seen, was clearly the era of the fastener. At its beginning there was the phasing out of dress-fasteners and dragonesque brooches, both of which still carried a few remnants of Celtic art patterns, in favour of other forms not of British origin. The turning-point appears to have come after the building of the Wall, when native cultures appear to have become more agrestic. Fasteners became purely functional, without much pretence at decoration. Many were mass-produced: only the umbonate disc-brooch, one of Nor'nour's specialities, appears to have carried on a tradition of enamelling, and even there the prototype pattern had been alien. Then there were penannular brooches. The penannular form was a clever safety-pin idea: the captive pin was pushed through the gap to achieve its safety-pin function by a subsequent part rotation of the hoop.

But penannular brooches were not new in the second century. In previous centuries there had been nasty little penannular brooches made of wire, quite small, and with terminals formed by the wire being bent back upon itself, for the purpose of keeping the pin captive. At the beginning they were not decorated. They were purely functional, and they were of common occurrence, many having been found on Roman sites, and on Romano-British sites as well as on native sites. At Maiden Castle, Wheeler[2] dated one of these brooches to the period AD 25–70. Of course, such a simple form of fastener could have been made by almost anybody who was prepared to bend a short length of wire. The brooches bore no national characteristics, though some had crimped terminals, and on others the little terminals were waisted. This was the first sign of a departure from the absolutely plain form, and the idea was repeated many times in varying localities. But towards the end of the second century a new form of penannular brooch suddenly made its appearance. It was larger, heavier, and its terminals had assumed a *zoomorphic* appearance. This form of penannular brooch is called zoomorphic because its terminals bear the features of head, snout, eyes and ears.

Here was the native Briton once more impressing his personality on what had been a thoroughly uninteresting form. Of course, zoomorphism had been known at various times in Celtic art, both on the Continent and in Britain. As an example, there is the animal head terminal on the bronze bangle from Snailwell (Fig. 62:1). Other examples of animal heads can be seen on the Iron Age bowl from Rose Ash, North Devon (Fig. 62:2) and on a bronze spur from Icklingham, Suffolk (Fig. 62:3). These are fairly realistic representations, in as much as one can appreciate them for what they are. On the other hand, the zoomorphic heads of the penannular brooches are abstract. This abstraction is essentially British, and northern, and it appears to have come about as a result of a combined effort on the parts of the Brigantes and the Votadini. The basic idea came from the terminal of

198

a Brigantian bangle (Fig. 62:4). Whereas, in the more naturalistic animal figures the face is always turned to the *outside*, in the case of the abstract or stylised form the animal faces *inwards*, the snouts being engaged in what Stevenson has described as a 'hoop-swallowing act'. The backs of the heads, which are on the outside, are always squared off.

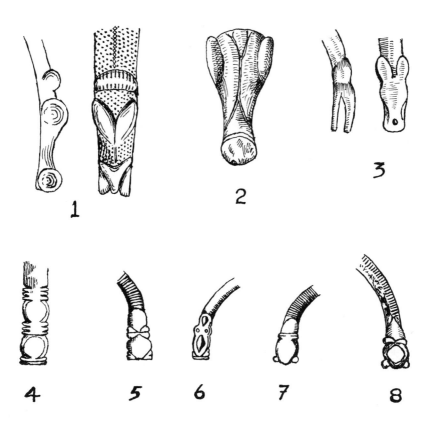

Fig. 62: *Zoomorphic Forms* 1. Terminal of Bronze Bangle, Snailwell, Cambs. 2. Head on Iron Age Bowl, Rose Ash, North Devon. 3. Terminal of Bronze Spur, Icklingham, Suffolk. *Terminals of Zoomorphic Penannular Brooches.* 4. Terminal of Brigantian Bangle, Isurium Brigantium. 5. Traprain Law. 6. Barnton, Midlothian. 7. Longfaugh, Midlothian. 8. Ford of Toome, Lough Neagh.

From all the evidence available, it would appear that the whole movement towards this new development was initiated by the Votadini, who were already making pins with rounded heads and a snout, though there were no eyes or ears. The snouts of these rounded pinheads are slightly upturned at the tip, which came about because the earliest pins had bent-back heads. One of these still exists at Traprain Law. The distribution of these proto-zoomorphic pins (as we must call them) is confined to Traprain Law, Newstead, and Covesea Cave. At Traprain Law they have been found in the fourth, third and second levels, suggesting a continuing history from the first to the third centuries. The Newstead evidence is more or less in agreement with regard to an early start, in that the pin came from an early building within the fort, and it is therefore Flavian, or at the latest Trajanic. The same proto-zoomorphic form made its appearance in Ireland, but there its length was doubled, with the top two inches being lavishly decorated.

Some of the Irish-made pins were exported to Britain, mostly to the Bristol Channel area. Boon[3] has reported on the discovery of two pins on Margam Beach, west Glamorgan, whilst another came from the Prysg Field, Caerleon, and a fourth from Silchester. In the Hadrian's Wall area a fifth pin (also Irish) was found at Cilurnum; whilst others came from Cirencester and Cassington, Oxon.[4] These finds are mentioned with but one purpose, and that is to show the universality of the proto-zoomorphic form, though its distribution in Britain is odd in as much as there are no occurrences of the proto-zoomorphic form except in the locations named. In effect, it was just another type of pin developed at Traprain Law, where the craftsmen were good at developing pins. Although the bent-back terminal brooch is known here, nobody thought to produce a larger brooch until the Brigantian bangle was seen (Fig. 62:4). It was lightweight because the metal was thin, and it was penannular, with surface mouldings at the terminals. Moreover, the ends were squared. The surface mouldings were separated from one another by transverse triple channelling: the transference of this terminal decoration to the proto-zoomorphic head resulted in the production of a genuine zoomorphic form. This coalescence of ideas could not have been long in coming, and the result was the appearance of a new brooch form, bangle size. Not all the steps in the evolution of this design are available to us, but some remain. For instance, the first terminal, Fig. 62:5, shows how the ears came into being. The double-end moulding is copied from the Brigantian bracelet, but it was unintentionally cut through during the process of head-shaping, when the metalworker filed from the end instead of from the side. This was sloppy workmanship; but the resulting division of the end moulding into two halves suggested their further shaping into ears, as indeed was the case, as can be seen in Fig. 62:7. Note that the moulding situated between head and snout still remains untouched. But by the time that Fig. 62:8 was made the eyes were made to match the ears. In all these cases the head is rounded, as it was on the proto-zoomorphic pins, and as the terminal is moulded in a similar manner to that of the Brigantian bracelet, for which a second-century date is claimed.[5]

Thus, by steady evolutionary methods did the zoomorphic form mature; and once stabilised it altered hardly at all. Most of the Scottish brooches were found within easy reach of Traprain Law, and some came from the oppidum itself. One was found in the third level, whilst another came from the upper 'native' levels. There are also zoomorphic pins from this site, with heads and terminals following closely the shape of those belonging to the brooches. Manufacture in the second half of the second century is attested by the occurrence of one brooch in the Roman and Romano-British levels, and perhaps the same is true of the second brooch; for, although it was found in the upper levels, its presence in the post-AD 196 context rather suggests survival. Some[6] would find it hard to accept this pre-196 evidence of a third-level find, preferring a date somewhere between the third and fifth centuries.[7] The association of the Longfaugh brooch with a Roman patera and buckle[8] is also discounted; yet small hoards of this nature were not uncommon around the end of the second century, because of the uncertainty of conditions after the abandonment of the Antonine Wall.

The brooches illustrated in Fig. 63 belong to the *Initial Form*, so called because they have resulted from primary development. Except for the ribbing on the hoops, all are plain. Ribbing also decorated both proto-zoomorphic and zoomorphic pins. The distribution of

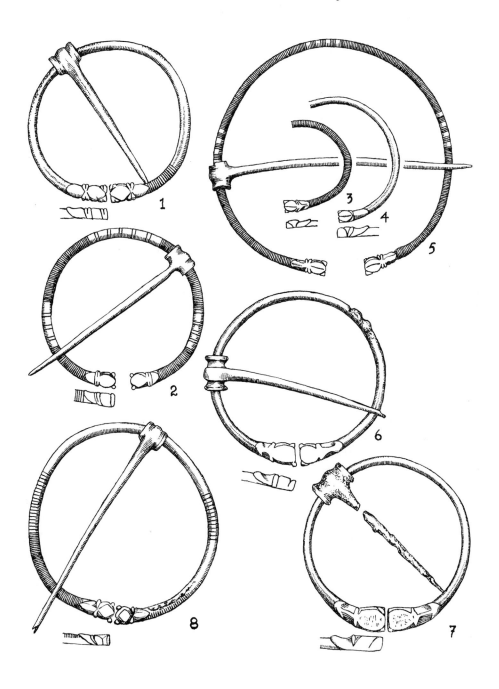

Fig. 63: Zoomorphic Penannular Brooches: Initial Form 1. 4. Traprain Law. 2. Longfaugh, Crichton, Midlothian. 3. Minchin Hole, Penard, Glamorgan. 5. Stratford-on-Avon, Warwickshire. 6. Aikerness Broch, Evie, Orkney. 7. Pinhoulland, Walls, Shetland. 8. Ford of Toome, Lough Neagh. (3/4)

zoomorphic penannular brooches seems to follow the same pattern as did the pins, but with one provision: none has been found in the Atlantic Province.[9] As in the case of the pins, there is an avoidance of Caledonia, but there is representation in Caithness, Orkney and Shetland. The Orkney and Shetland brooches are the latest in the British series (Fig. 63:6 and 7), but these cannot be dated by association. Several bits and pieces of Roman objects, often pottery, have been found in Orkney, mostly in the brochs, and since datable objects amongst these bits and pieces belong to the latter end of the second century, they (together with the brooches) could have been carried north after the sack of Traprain Law, where they had been made: alternatively, they could have been acquired in the course of trade, seeing that both the Orkney chiefs and the Votadini were friendly to Rome.

Products from Ireland and the north found a point of ingress on the north side of the Bristol Channel. Here, at Minchin Hole, Penard, Glamorgan,[10] and at Stratford-on-Avon,[11] two brooches which are alike, except in size, were found (Fig. 63:3 and 5). Both are early examples of the Initial Form; and especially with regard to the eyes, may be compared with Fig. 63:1, from Traprain Law. The medial line on the head is a feature also found at Traprain Law (Fig. 63:4), so that on both form and decoration a northern (Traprain Law) source is suggested for the Minchin Hole and Stratford brooches. The discovery of these northern brooches so far south has produced some odd reactions, causing Savory[12] to insist on a southern origin for the Initial Form, and then not before the fourth century. As reason for his insistence, Savory points to the association of the Minchin Hole brooch with fourth-century finds. No allowance is made for the second-century Scottish evidence, or for the fact that a brooch was found at Segontium, Caernarvon,[13] and along with others of a pseudo-zoomorphic nature, it can be put into a period prior to AD 200.[14] Clear cases of survival are numerous, relating to personal items like brooches and pins: a brooch, worn as an armlet, was found in a Saxon grave at Bifrons, Canterbury.[15] Should anyone doubt the feasibility of Traprain Law brooches being able to reach south Wales, he should look back at the distribution map of dress-fasteners. Irish proto-zoomorphic pins reached the same area, as we have seen. The chief objective appears to have been the Silurian market.

Mention was made above of pseudo-zoomorphic brooches. These are small brooches, of approximately the same size as the bent-back terminal brooch, and sometimes, like it, made from wire with crimped-back terminals. Pseudo-zoomorphic brooches are distinguished by the fact that the bent-back terminal — this brooch is often cast as well — is rudely fashioned in imitation of the true zoomorphic form. Brooches from Segontium,[16] Birdoswald,[17] and South Shields[18] have terminals which have achieved a credible representation of the zoomorphic form. Some makers of these little brooches came very near to understanding the form, whilst others did not, but their attempts at copying show no evolutionary sequence at all. All the same, they are useful adjuncts for dating purposes, in that some are closely dated, more so perhaps than with the genuine zoomorphic form. No less than seven pseudo-zoomorphic brooches have medial lines on the terminal heads, reflecting the occurrence of these lines on the Minchin Hole and Stratford brooches, as well as the brooch and pin from Traprain Law. Therefore the suggestion is that copying began at an early stage in development. This suggestion becomes fact at Segontium. George Boon[19] writes that possibly reliable associations here date at least one of these brooches to some period before AD 200. This date is very pertinent to the history of the Initial Form, and of course indirectly it confirms the Traprain Law and Longfaugh evidence. Valid support for the early dating is forthcoming from far-away Orkney. A brooch from the broch of Okstrow, Birsay,[20] was associated with an occupation that also produced three pieces of a thick Sigillata bowl, type D. 45. This bowl has an upright rim, and normally a lion-head mouth-piece, and belongs probably to the latter part of the second century.[21] Pseudo-zoomorphic brooches thus reflect what was going on by way of development of the Initial Form; and their own second-century date should make it easier to accept a similar date for the genuine zoomorphic type.

In Britain, it is clear that development was halted by the events of AD 196, when the workshops were destroyed; for which reason brooch design never progressed beyond the Initial Form. The terminals remained too small for applied decoration. The credit for decorating the terminals goes to Irish craftsmen, who acquired the Initial Form, and at first faithfully copied it. But soon, in Irish hands, the terminal size began to increase, and as terminals became larger so was decoration added, and somewhat later decoration was enamel backed. Large numbers of these brooches were produced in Ireland, and the form became distinctly Irish. Working in the midst of an heroic society, the metalworkers had every encouragement to produce striking brooches; which they did, and some of the specimens they produced will be discussed in the following chapter.

NOTES

1. Production between the second and the fifth centuries was prolific; and because of the large numbers remaining, they have had to be made the subject of a separate monograph. The chapter which follows does nothing more than summarise the development of the Initial Form, as set down in the monograph. H.E. Kilbride-Jones, *Soc. Antiq. Research Comm. Report*, XXXIX.

2. Wheeler, 'Maiden Castle', p. 264, Fig. 86:7.

3. G.C. Boon, 'Two Celtic Pins from Margam Beach, west Glamorgan', *Antiq. Journ.*, LIV (1975), p. 400.

4. E. Fowler, 'Celtic Metalwork of the Fifth and Sixth Centuries A.D.', *Arch. Journ.*, CXX (1963), p. 121, Fig. 2.

5. M.V. Jones, 'Aldborough, West Riding: Excavations at the South Gate and Bastions, and at extra mural Sites', *Yorks. Arch. Journ.*, 43 (1971), p. 39, Fig. 22:20.

6. H.N. Savory, 'Some sub-Romano-British Brooches from south Wales' in Harden (ed.), *Dark Age Britain*, pp. 40–58. Also, Fowler[4], p. 99, and L. Alcock, *Arthur's Britain*, pp. 236, 261.

7. Savory[6], pp. 40–58.

8. *Proc. Soc. Antiq. Scot.*, II (1859), p. 237.

Ibid., V (1865), p. 188.

9. Excepting, of course, a pseudo-zoomorphic brooch found in North Uist.

10. Savory[6], pl. V:b.

11. *Proc. Soc. Antiq.*[2], XXVII (1914–15), p. 96, Fig. 3.

12. Savory[6], pp. 40 ff.

13. R.E.M. Wheeler, 'Segontium, and the Roman Occupation of Wales', *Y Cymmrodor*, XXXIII (1923), Fig. 58:6.

14. Mentioned by G.C. Boon in *Arch. Camb.*, CXXIV (1975), p. 65, and note 35, and later dated by him to before 200.

15. *Arch. Cant.*, X (1876), Fig. opp. p. 303.

16. Information kindly supplied by G.C. Boon.

17. *Trans. Cumb. & West. Ant. & Arch. Soc.*, n.s., XXX (1930), p. 170.

18. *Arch. Ael.*[4], 11 (1934), p. 97, Fig. i and 2.

19. In a letter to the author.

20. 'Donations to the Museum', *Proc. Soc. Antiq. Scot.*, XI (1876), p. 85.

21. Curle, 'Objects of Roman and Provincial Roman Origin found in Scotland', p. 285.

23. Craftsmanship in Ireland (II)

In this continuing story of Irish craftsmanship, consideration has already been given to a class of decoration which appears on several rather spectacular objects; but it petered out some time after the period represented by the Roman Conquest of Britain, and with it went a very advanced technique in applied decoration. The importance of some of those patterns to the continuing argument is that some, as with that on the Bann disc, are triskele based, whilst others relied on the simplified palmette. In this chapter, the story is taken up again; and whilst the later period with which it deals, as with the corresponding period in Britain, is characterised by the appearance of numbers of small objects of a personal nature, nevertheless both these prime motifs are retained during the greater part of the period covered. Nevertheless, the triskele will outlast the palmette in Ireland, since all triple spiral patterns are based on the triskele. The chief objects for consideration will be zoomorphic penannular brooches, handpins, latchets and the like — all of them fasteners intended for securing personal attire. As with Britain, even in Ireland it was an age of fasteners.

But palmettes and triskeles of themselves did not make up the entire decoration of the period, which runs parallel with but outlives the Roman Occupation of Britain. Other patterns are also noted, and these are basically Romano-British. Each pattern will be discussed as and when the object upon which it appears itself comes up for consideration. In the meantime it is necessary to start at the beginning and to note that a cross-Irish Sea connection is testified by the presence in Ireland of Traprain Law devised forms, particularly the Initial Form of the zoomorphic penannular brooch series. The Irish became acquainted with this Initial Form in the early days of its development, as evidenced by the discovery at Toome[1] of a basic form of brooch (Fig. 63:8). There is nothing in the distribution of these brooches which would indicate how they reached Ireland; but the probability is that trade was conducted through Galloway. The Irish took serious note of the new form, so that, in a very short time, zoomorphic penannular brooches similar to these were being made in Ireland. From the map, Fig. 64, it will be noted that these Irish-made brooches have been found over much of the country, though there is a slight western emphasis, in that the majority occur along the Shannon or to the west of it. Again, like their British counterparts, many of these Irish brooches are plain, whilst others have ribbed hoops; and the ribbing, normally divided up into four distinct zones, follows closely upon that seen on the Longfaugh and Stratford-on-Avon brooches (Fig. 63:2 and 5), so that, whilst Fig. 65:1 is plain, Fig. 65:2, 3 and 4 have ribbed hoops, in the British manner. Even the medial lines on the heads of the Stratford-on-Avon, Minchin Hole and Traprain Law brooches have been repeated on Fig. 65:2, a specimen which comes from Roosky, Co. Roscommon.[2] But the Irish lost little time in applying patterns

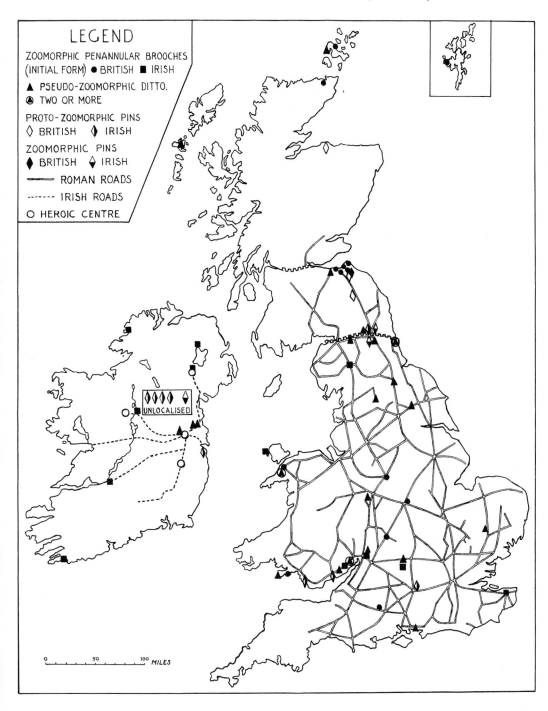

Fig. 64: Map Showing Distribution of Zoomorphic and Pseudo-zoomorphic Penannular Brooches, and of Zoomorphic and Proto-zoomorphic Pins.

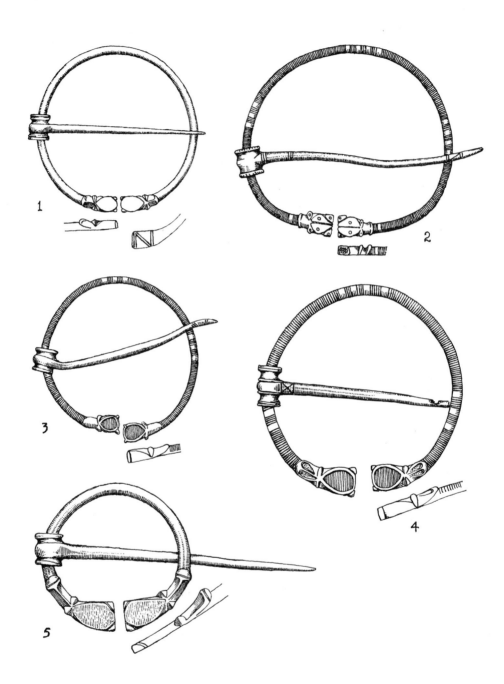

Fig. 65: *Irish Zoomorphic Penannular Brooches: Initial Form* 1. Kirkby Thore, Cumbria. 2. Roosky, Co. Roscommon. 3. Abingdon, Berkshire. 4. Armagh. *Group A_1* 5. Toomullin, Co. Clare. (3/4)

to their brooches: the reverse sides of the terminals of Fig. 65:1, which was found at Kirkby Thore, Cumbria,[3] have N-patterns inscribed upon them, whilst the pinhead has been improved by the addition of mouldings. This N-pattern will persist, and it will be there, on and off, for the greater part of the existence of the brooch, as a type, in Ireland. It is not peculiar to any one series.

Therefore, right at the beginning, there was a basic desire to make the brooches more attractive. Amongst the Initial Form brooches there were small additions to the pins, such as the transverse lines on the pin of Fig. 65:2, along with the little nicks on the outer mouldings of the pinhead. The terminals have hollow punch marks upon them. But enamelling also entered the field of decoration: the brooch from Abingdon[4] (Fig. 65:3) and the brooch from Armagh[5] (Fig. 65:4) both have enamelled terminals, in addition to decorated pins. Note the improved mouldings of the pinheads: that of the Armagh brooch is particularly fine.

These were small beginnings. However, it was inevitable that somebody would draw the line between decorated and undecorated brooches. What results is a new division of Irish zoomorphic penannular brooches into four main groups, all of which are descendants of the Initial Form. These are Groups A, B, C and D. Because of the existence of recognisable variations within each Group, except Group D, then it is possible to have sub-groups within the main groups, each sub-group being made up of brooches with the same manufacturing characteristics, like a peculiarity or a pyramidal eye. Such characteristics are traceable to the work of a single metalworker, or alternatively to one workshop. Overall, form is paramount, and also it is variable: so that, whilst sub-groups are characterised by certain small peculiarities, it is the variations in form that determine the inclusion or otherwise of brooches into one group or another. Group C embraces the largest number, because Group C brooches were made during the most productive period in the history of the form. Groupings must never be attempted by reference to decoration alone, for decoration is ancillary. Decorative motifs can be anybody's choice, and indeed they are, and they can be borrowed from anywhere. Form, on the other hand, is more in the nature of a personal preference for line, so it can be compared with the lump of clay in a potter's hands. But the craftsman's attitude to decoration is perhaps more conservative and therefore less personal. Yet, withal, decoration is often more valuable for dating purposes.

The drawing of the line between undecorated and decorated brooches leads to the isolation of Group A, which is made up entirely of plain specimens. Brooches of the remaining groups are decorated. However, because all these groups are made up of descendants of the Initial Form of brooch, its life was correspondingly short in Ireland, judged on the number of extant specimens. The plain brooches of Group A can be divided into two distinguishable series, the feature that separates one from the other being eye shape. In the A_1 brooches the eyes are pyramidal, whilst in sub-group A_2 the eyes are rounded. There is little deviation from these eye-shapes, and the implication is that there were two workshops engaged in turning out brooches. The first workshop appears to have been situated in central Ireland, whereas the second one was situated nearer to the east coast. There is nothing remarkable about this: each served its own community. The earliest brooches belonging to series A_1 have points in common with the Initial Form. A developed specimen[6] is shown in Fig. 65:5. A_1 brooches are probably of third-century-AD manufacture. On the other hand, brooches included in the A_2 series are later, perhaps of fourth-century date; so that the A_2 workshop may have begun

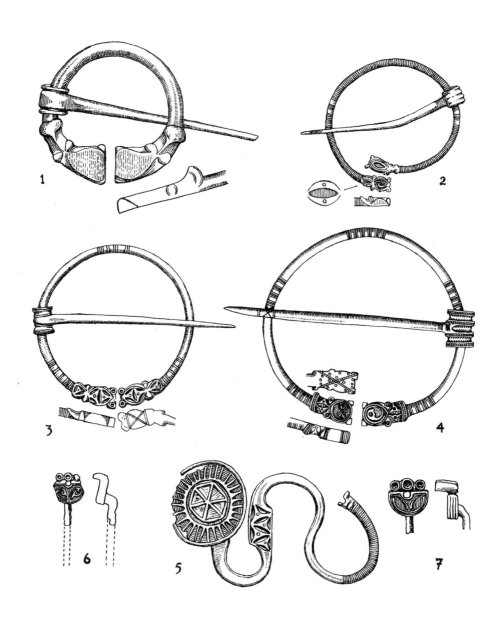

Fig '6: *Irish Zoomorphic Penannular Brooches Group A₂*: 1. Co. Waterford. *Group B₁*: 2. Knowth, Co. Meath. 3. River Greese, Co. Kildare. 4. No Locality, Ireland. *Irish Decorated Bronze Latchet*: 5. Newry, Co. Down. *Three-fingered Handpins*: 6. Culbin Sands, Morayshire. 7. F.J. Robb Coll., 546.37. (3/4)

manufacture only after the first workshop had ceased production. The A_2 brooches are not so refined; and in addition to the rounded eyes, there is a pronounced up-turn to the snout-tip (Fig. 66:1). These characteristics are common to the entire sub-series.

Decorated brooches present a different picture, because there are wider implications, and these relate chiefly to the form of decoration which each bears. The earliest decorated brooches are undoubtedly those of Group B. But here again there are two sub-groups, B_1 and B_2. The first of these is made up of the earliest brooches. An early example of a B_1 brooch is a specimen from Knowth[7] (Fig. 66:2). Basically, there is little difference between its form and the early brooches of the Initial Form, with its closely ribbed hoop divided into three zones; but there is a change in the pinhead shape. This was experimental, and the new form did not last very long. But the brooch form overall led to the production of some quite sophisticated types, like Fig. 66:4. But in the case of the Knowth brooch, attention should be given to the very interesting decoration on the snout, for this miniaturised pattern has far reaching implications. An enlargement has been made, in order to show its direct analogy with the more advanced pattern on the handle of the Nijmegen mirror (Fig. 21:2) and some scabbard decoration (Fig. 32:3). Judged on form alone, the Knowth brooch is early, perhaps dating from the closing years of the second century, and this date immediately brings to light a time-lag, here of at least a century's duration, maybe longer. This time-lag is most disturbing, but it has to be taken into account when dating Irish objects. Later, the gap widens further. The time-lag factor appears to increase with the length of the Occupation in Britain; whereas, before the Invasion, there could have been no more than a few years involved.

In Ireland, pin length serves as a rough guide to date; so that it can be said that the shorter the pin the earlier is the brooch. The Irish favoured a long pin, but it did not materialise all at once: the increase was a gradual process. The proto-zoomorphic pins were similarly lengthened, and some are up to 14 inches in length. Later brooches have pins which are roughly, in length, double the diameter of the hoop. Other Irish eccentricities are limited to pinhead treatment and the addition to the pin, just above its point, of diagonal or transverse lines. The maker of B_1 brooches must have been quite short-sighted, since the patterns are small, almost minute, and they were made to fit into limited spaces. If the double spiral on the terminal of the Knowth brooch looks a bit messy, this is because its maker tried to get too much into too small a space; yet, curiously, a somewhat similar messy pattern can be seen on the three-fingered small handpin from the Culbin Sands, Morayshire (Fig. 66:6). This might be only a coincidence; but it must be remembered that the Deskford boar's head, Irish work beyond doubt, came from the next county. Another handpin in the same style is in the Robb Collection in the Ulster Museum (Fig. 66:7).

Linear perspectives broaden when the River Greese, Co. Kildare, brooch (Fig. 66:3) is considered. The main terminal decoration here is made up of paired spherical triangles. Paired spherical triangles with dots at their centres occur nowhere but in Ireland. These dots are an Irish eccentricity based on a misunderstanding of the pattern on the snout of the Knowth brooch. On this River Greese specimen, spherical triangles occur in profusion on terminals as well as snouts, as though they were new and everybody was delighted with them. Their liberal use here is reminiscent of the usage of spherical triangles on the Newry latchets, one of which is shown in Fig. 66:5. In profile, the River Greese brooch has the same form as has the Knowth brooch, and for that reason both may belong to the third century's early years, with the latter perhaps belonging to the closing years of the second. On the reverse sides of the terminals of Fig. 66:3 there are saltire-type patterns.

The transference across the Irish Sea of the Nijmegen mirror-handle pattern is not a lone occurrence: another relates to the pattern on the disc of the Newry latchet. The Newry latchet discs have, at their centres, emasculated versions of the hexafoil motif; and as a background to these there is the sunburst pattern. Hexafoil motifs should have six petals radiating from a common centre, the outer tips being conjoined by a further six petals. But here at Newry the petals are not petals at all, but straight cut representations of what is meant to be petals. This crude form of representation was much favoured by the Romans, and they put this emasculated form on their altars, their tombstones, and their antefixes. Examples have been given above,[8] and on balance they are of second-century occurrence, with a persistence into the third. On the other hand, the sunburst pattern is earlier, appearing on first-century bowls and dress-fasteners,[9] and in stylised form it is common on second-century disc-brooches. However, at Newry, representation of the sunburst pattern is in its early naturalistic form, so that one wonders where it had been kept in reserve between the first and the third centuries. This is the sort of occurrence in Ireland that makes dating a more or less impossible undertaking when assessing Irish products.

Such alien patterns and motifs may have been in more common use in Ireland than is generally supposed. There is the case of the metalworker from Dooey, Co. Donegal,[10] whose antler trial piece (Fig. 67:1) bears this same emasculated version of the hexafoil motif, together with multiple-lozenge patterns and the chevron pattern, the only Celtic patterns present being spirals. It might be thought odd that a metalworker living and working on the north-west coast of Ireland should have favoured patterns of alien origin; but the fact is that he did, and presumably some found their way on to the metalwork of the times. The paucity of metal objects is a drawback here, and prevents a better understanding of the situation.

In Britain, multiple-lozenge patterns were much favoured by metalworkers, especially as decoration for dragonesque brooches, thus pointing to a late-first- or early-second-century usage. The Dooey trial piece cannot be closely dated, although, according to the excavators' account of the excavation, it came from a sealed layer which belonged to 'the early centuries AD'. Maybe the Irish learned about these patterns as a result of their raids on Britain: either that, or communications were better than is generally imagined. There are, of course, Roman finds in Ireland, and those of second-century date can be equated with the disaster of 196, when Britain was left virtually defenceless as a result of Clodius Albinus' expedition to Gaul in his attempt to gain the purple. It is known[11] that the Irish Sea was a highway for the Irish, for they continued to paddle their canoes up and down, more or less at will, all the time causing trouble by making sporadic raids on the British coastal areas, in their search for booty. These conditions led to the establishment *c.* 275 of an Irish aristocracy in Pembroke. Apart from that, the hippocampic heads seen on the Petrie Crown indicate contact with Brigantian metalworkers in the early second century; and a bit of two-way business might be proved by the discovery at Kirkby Thore of an Irish zoomorphic penannular brooch (Fig. 65:1) and of proto-zoormorphic and zoomorphic pins, also Irish, at Cilurnum and Hunnum. At this stage of our knowledge, it is still difficult to decide exactly what was going on; yet evidence of contact is proved.

Another B_1 brooch, unlocalised (Fig. 66:4) has snouts decorated with paired spherical triangles; but amongst the head decoration there is the first appearance of ornithomorphic

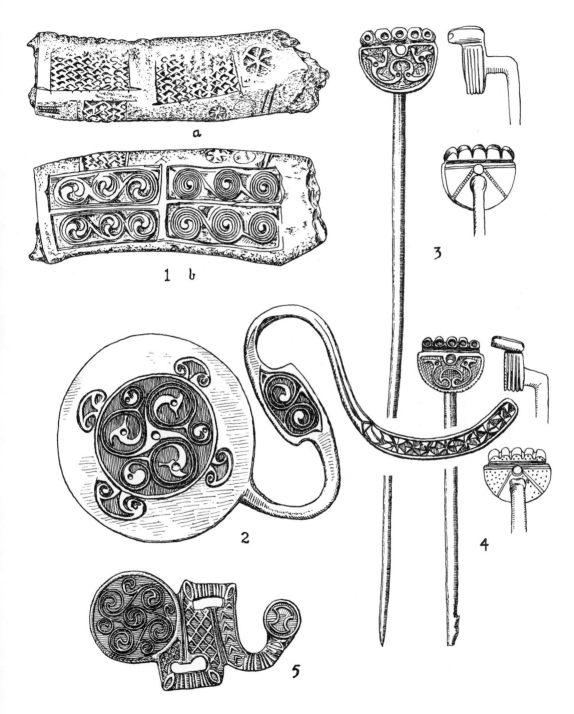

Fig. 67: 1. Antler Trial Piece, Dooey, Co. Donegal. 2. Decorated Bronze Latchet, Ireland.
3. 4. Enamelled Handpins, Ireland. 5. Enamelled Bronze Latchet, Dowris, Co. Offaly. (1/2) (9/8) (3/4) (3/4)

finials. Like most whims, this fashion did not last long, but it is approximately datable. Ornithomorphic finials like these appear on an Irish decorated latchet (Fig. 67:2). This latchet is without provenance. The general pattern on the disc is based on the triskele, and the hippocampic finials are similar to those on the Petrie Crown. Clearly, finials treated in this manner were in fashion; and here contemporaneity is assured for both hippocampic and ornithomorphic finials, both styles occurring on this disc. The overall impression left by the decoration on this latchet is that in some ways it represents a follow-up to that seen on the Bann disc. The overdependence on elongated trumpets was to be relaxed, leaving a basic theme for other designers to do something with. And here we find the result of somebody's efforts on the disc of this latchet. It is quite clear that this form of triskele, with its inturned spiral-like ends, was the basis for many later designs, such as those appearing on the B_2 brooches and on the prints and escutcheons of the hanging-bowls. In Ireland, there was always a tendency for finials to be swollen: good examples can be seen on the Dooey trial piece. Another later tendency was to hollow out the swollen finial, the result being that spirals like those on the Dowris, Co. Offaly, latchet (Fig. 67:5) came into being. These hollowed-out finials are often mistaken for double spirals: these and genuine double spirals are both represented on the disc of the Dowris latchet.

The hippocampic finials of Fig. 67:2 are not a lone representation: the same type can be seen on two handpins (Fig. 67:3 and 4), both unlocalised, though found in Ireland,[12] and both clearly the products of the same workshop. This type of finial cannot be seen on any of the zoomorphic penannular brooches. But, if we want to date the latchet and the handpins, the dual representation of hippocampic and ornithomorphic finials is very pertinent. Ornithomorphic finials are very common on proto-handpins (Fig. 68), all of which are the products of a single workshop, since the family likeness is obvious, and likewise the techniques employed. Now, the first of these pins, Fig. 68:2, was found at Oldcroft, Gloucestershire,[13] not far from Lydney; and this find is of great importance, in that this pin is the only one with a *terminus ante quem* dating, in this case of 354—59. This pin was found in a hoard of 3,330 bronze coins, eleven of which dated before 330, whilst the latest was one of Julian (354—9). Yet the pinhead decoration consists of a much stylised version of the palmette, with a lobe to each side, each lobe being of an ornithomorphic character. These ornithomorphic features are precisely similar to those of the Fig. 67:2 latchet, and the B_1 brooch, Fig. 66:4, with which this discussion began. But if we relate these to the Oldcroft pin, then we make them too late; so we must reconsider the Oldcroft pin for more information. Its decoration is composed of a central highly stylised palmette flanked by two lobes, and in this style we note a direct affinity with the style of decoration on some early Occupation terrets in Britain, notably those found at Westhall (Fig. 27:2 and 3). The Westhall terrets, as most people will agree, were buried about the time of the Boudiccan uprising in AD 60, but this leaves a time-lag of some 300 years, a figure that is totally unacceptable. But a long survival period is indicated for the pin, since it is made of silver, and it was considered valuable, and that is why it was buried with the coins. Due allowance must be made for its sentimental value: a period of perhaps a century, or even a century-and-a-half may not be too long an allowance, thus making the Oldcroft pin old, by the standards of the day, when it was buried. Devolved patterns, imitating stylised decoration current in the mid first century (for the terrets), were probably known in Ireland in the mid to late second century, due allowance being made for the time-lag factor. By taking into account every known factor relating to the Oldcroft pin, it now seems reasonable to assume that it was made at the beginning of the third century, or at the latest by the middle of that century. This date fits in well with the estimated dates of the B_1 brooches. But it means very little when we talk of hippocampic finials, since these were still in vogue as late as the Hiberno-Norse period (see Fig. 76:6). Thus we are no nearer to a suitable date for the handpins (Fig. 67:3 and 4), though the style would make them not much later than the latchet under discussion.

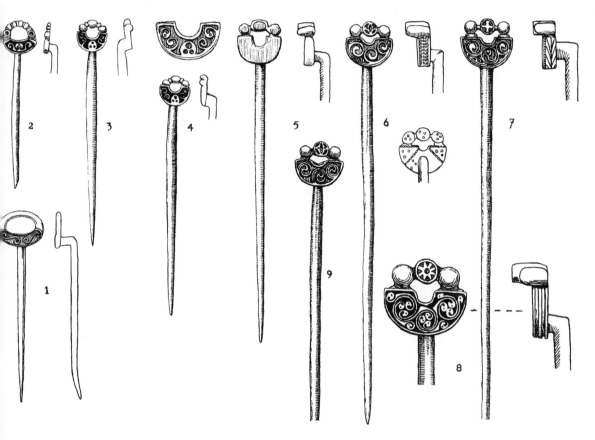

Fig. 68: *Pin, with Decorated Oval Pinhead*: 1. Hunnum, Hadrians Wall. *Proto-Handpins*: 2. Oldpark, Gloucestershire. 3. Ireland. 4. 6. Castletown, Kilberry, Co. Meath. 5. Moresby, Cumbria. 7. Norries Law, Fife. 8. Gaulcross, Morayshire. 9. Clonmacnoise, Co. Offaly. (3/4)

Apparently, spirals took over from these fancy finial representations, though a remarkable survival of the fashion is noted in the case of the Gaulcross pin[14] (Fig. 68:8), on which spirals with ornithomorphic finials appear, lavishly displayed. But the remaining pins demonstrate how the spiral took over; so that on pins, Fig. 68:6 and 7, spirals with swollen finials have become the standard form of representation. The Irish swollen finial could be a simplified form of the ornithomorphic type: there is small reason why this should not have been the case, since, by the addition of a dot, one would again have a fair ornithomorphic representation. This course of events is noted in the case of the two pins found at Castletown, Kilberry, Co. Meath (Fig. 68:4 and 6). Finds such as these turn speculation into fact. The smaller of the two Castletown pins is identical with Fig. 68:3, from Ireland, though the former is of silver. Pins with short stems are generally early, the pin length becoming greater with the passage of time. The fact that this small Castletown pin was found with another nearly twice its length is another clear case of survival over a long period: it is also clear evidence of a change of decoration; for between the two periods of manufacture, palmettes and ornithomorphic finials had been dropped in favour of single spirals with slightly swollen finials.

Cross-channel relationships again come to the fore, for our consideration. Nobody can deny that Fig. 68:7 is very like the larger of the two Castletown pins; yet the former comes from Norries Law, Fife. On its reverse side there are the same hollow punch marks as exist on the Castletown pin. The two are analogous. The Norries Law pin was found in association with a sub-oval silver plaque having three (originally four) heavily embossed coils, resembling those on the so-called sacrificial dishes (Fig. 5:2). The embossed technique is identical, the form is very similar. Irish elongated trumpets are worked into this embossed pattern on the plaque, and with all these points considered together there is sufficient evidence to dispel any illusions about its being 'proto-Pictish material'.[15] The plaque may be of second-century date, but the pin is a full 100 years later. The character of the Gaulcross pin puts it into the same school of art. Stevenson[16] has already remarked on the similarity between the decoration on the Gaulcross and Norries Law pins and that of what he describes as the best of the Irish latchets. Views such as his should make it easier to accept Irish influence amongst the decoration on the massive armlets; and there can now be less doubt about the origin of the Deskford boar's head.

Still on the same subject, next there is the pin found at Port Moresby (Fig. 68:5), and though plain, it must be regarded as belonging to the same family. It was found near the trail followed by the Mealsgate and Kirkby Thore brooches as well as the Cilurnam and Hunnum pins. The Oldcroft pin came from the territory of the Silures, to whom the Irish sent their proto-zoomorphic pins, found at Margam Beach and Caerleon. The later elaborate Celtic temple at Lydney was dedicated to the god Nodens, the only dedication to this god in Britain. Nodens was an Irish god. With the passage of time, more and more details are being added to this picture of Irish influence in Britain; and the whole situation is important in that it makes it easier to decide whence came the inspiration for the Artistic Revival, just before and after the Withdrawal.

The pins of Fig. 68 and the B_1 brooches have shown how it was possible, even with the help of primitive tools, to apply minute decoration in a small space, which must have necessitated a steady hand and short eyesight. For all we know, brooches and pins may have come from the same workshop, since the only localised pins came from the Meath area, which was also the centre of brooch manufacture, for the B_1 sub-group. The pins tell us that this factory was in existence in the third century. In the following century, another workshop was established somewhere near Athlone, or even a little farther to the west, perhaps on 'the old plain of the Soghain'. This workshop was responsible for the B_2 brooches, large specimens, upon which there were spiral patterns in the best tradition of Celtic art (Fig. 70). This persistence of the art in the west of Ireland was one of the most encouraging events of the times, since elsewhere the art was in a sorry state. Of Britain, Leeds[17] wrote:

it is an uncontestable fact that from about 250 A.D. to the end of the Roman period, objects other than pottery, which exhibit in any degree the influence of Celtic Art in their decoration, are distinguished by nothing so much as their scarcity.

In Britain, the art had already gone underground; but in Ireland, in Soghain territory west of the Shannon, the art was persisting, and it was being carried on with vigour, for workmanship was first-rate, and enamelling was used as a background to the patterns. But the B_2 makers relied heavily on the spiral for decoration, as seemingly the only motif left to them, apart from a knowledge of the triskele as a basis for the layouts. Thus the continuing influence of the triskele was still being felt, and it continued to persist. However, the general approach to brooch-making by the B_2 workshop was different in most respects, and certainly different from that of the B_1 series, as may be seen by reference to the B_2 brooches illustrated in Fig. 70.

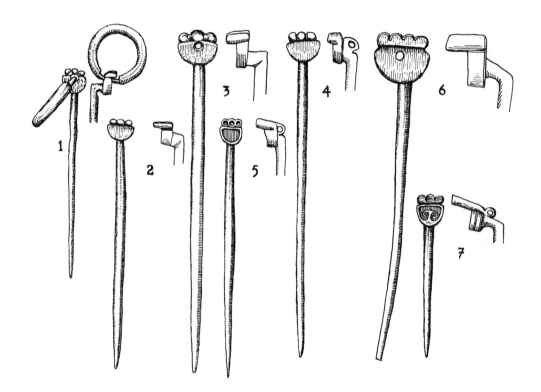

Fig. 69: Three-fingered Handpins 1. Ireland (Buick Coll., U.M. A186-1966). 2. Co. Westmeath. 3. Ireland. 4. Enniskillen, Co. Fermanagh. 5. Laragh, Enniskillen, Co. Fermanagh. 6. Maghera, Loughros Beg, Co. Donegal. 7. Ballykinvara Townland, Co. Clare. (3/4)

As we have seen, three-fingered handpins were developed at Traprain Law, where one occurred in the 'native' levels, after the reoccupation of the oppidum after the disaster of 196. As with the Initial Form of the zoomorphic penannular brooch, the three-fingered handpin reached Ireland, and one (Fig. 69:1) from Ireland in the Ulster Museum, is identical with the Traprain Law specimen, Fig. 60:9. Once it had reached Ireland, the form underwent remarkably little variation. It remained plain until enamel and a loop at the back were added in the later years of its history here. A selection of Irish three-fingered handpins is shown in Fig. 69, and from this it will be seen that the pins never achieved the lengths attained by the five-finger variety.

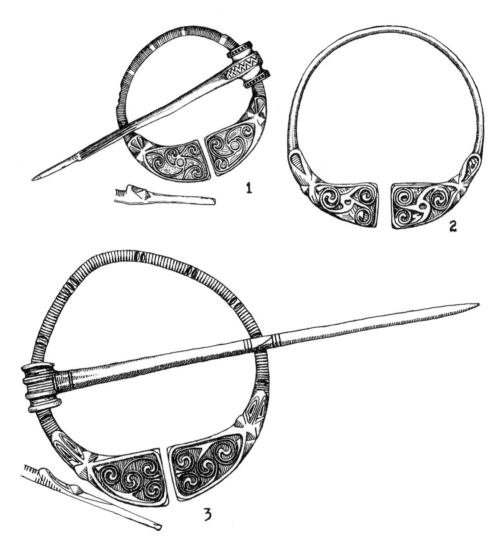

Fig. 70: Irish Enamelled Zoomorphic Penannular Brooches: Group B$_2$ 1. 2. Ireland.
3. near Athlone, Co. Westmeath.

Zoomorphic penannular brooches of Group C are predominantly eastern in Ireland, not more than two brooches having been found west of the Shannon. Group C numerically is also the largest group. Many of the brooches are expertly made, robust, often impressive, but the decoration is sometimes intangible for the reason that it is based on

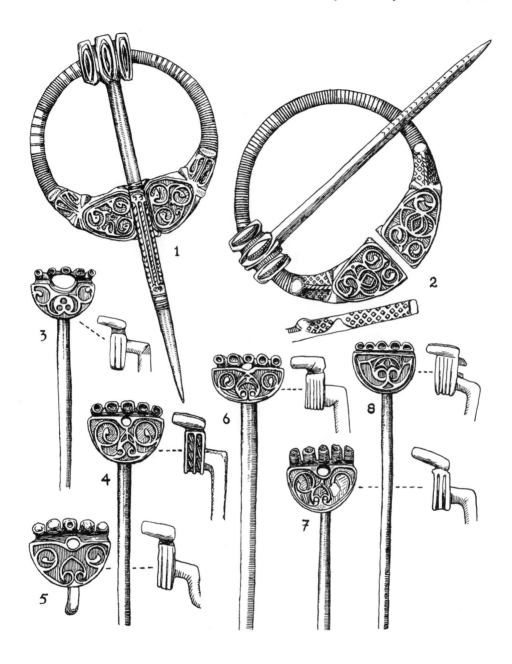

Fig. 71: *Irish Enamelled Zoomorphic Penannular Brooches: Group C_1*: 1. 2. Ireland.
Irish Enamelled Five-fingered Handpins 3. Clough, Co. Antrim. 4. Portglenone, Co. Antrim. 5. Ireland.
6. Ireland. 7. Brighton Down, Sussex. 8. Clougher, Co. Tyrone. (3/4)

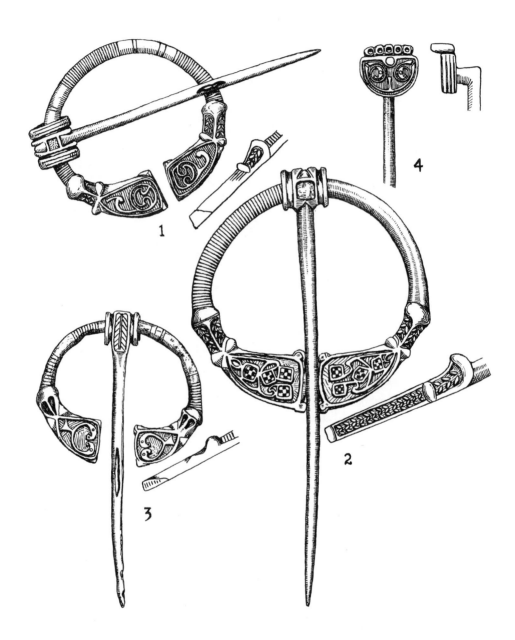

Fig. 72: *Irish Enamelled Zoomorphic Penannular Brooches Group C₂*: 1. Arthurstown, Co. Kildare.
2. Armagh. *Group C₃*: 3. Bloomfield, Co. Roscommon. *Five-fingered Enamelled Handpin*
4. Ireland (W.194). (3/4)

no known artistic tradition. All designs were borrowed or were imitative; but they were utilised in an interesting and sometimes striking manner, but clearly without understanding. Two brooches (both without locality) belonging to the C_1 sub-group are shown in Fig. 71:1 and 2. These are very ornate specimens, even showy, with enamel-backed designs on the terminals. These designs are clearly inspired by those triskele-based spiral designs on the B_2 brooches. The disingenuous spirals with their swollen finials have parallels on some of the handpins, six of these being shown in Fig. 71. These must be from the same workshop, and perhaps they were made about the same time as the brooches: but there is an additional motif, in that the highly stylised palmette is included as a centre-piece. The form of this palmette in Fig. 71:3, together with the general style of the pinhead, indicate descent from similar palmettes on the proto-handpins of Fig. 68.

Possibly the most interesting feature of the C_1 brooches is in the form of their pinheads, which are divided up into three identical sections, each ornamented with a seed- or petal-like pattern in false relief. This treatment of the pinhead is entirely new, and if clumsy is nevertheless ingenious and very striking.

Pinheads of the C_2 sub-group are equally big, but the treatment of them is rather different (Fig. 72:1 and 2). On these attention is focused on a small square of enamel which appears at the centre. But these C_2 brooches are clearly later, for now they are being made in an era which is characterised by the use of millefiori enamel. For that reason, decoration on the whole is less elaborate, as invariably happened once millefiori became popular: for its very popularity helped to kill off all forms of relief decoration and linear patterns. Thus, the advent of millefiori was about the worst event in the history of the art, for it heralded a decline in skills, with a corresponding decline in craftsmanship. The application of millefiori enamel to the terminals of brooches was a simple procedure, occupying few man-hours, and for that reason it pushed aside all the old art forms.

Brooches of the C_1 sub-group were distributed on the north and south sides of Lough Neagh; but the C_2 sub-group had a Kildare-Westmeath distribution, though one specimen got to Mealsgate in Cumbria. But the C_3 sub-group brooches have a south of Lough Neagh distribution, but are contained by the Shannon on the west, and by an imaginary line drawn from Dublin Bay to Athlone. Perhaps the workshop was located in the Cavan region. C_3 brooches are more devolved, both artistically and in the matter of form. Decoration begins with simple devolved patterns on enamel backgrounds, and ends with millefiori enamel (Figs. 72:3 and 73:1). But something was beginning to go wrong. Clearly, these brooches cannot be earlier than the fourth century, and some may even belong to the fifth. Now this was a period of increasing pressure by the Irish on Britain, by the stepping-up of raids, and this could have made for increasing affluence in Ireland, since the parasites were doing well, as the increase in the number of coins of this period indicates. But affluence does not of necessity bring about an improvement in craftsmanship, but in fact the reverse. The mass-production of second-century trumpet brooches is a case in point. The only improvement is in numbers. Mass-production leaves less time for refinements. So the numerically large C Group of zoomorphic penannular brooches becomes a reflection of the events of the fourth century. The C_4 sub-group brooches are typical productions of this period, since the workmanship is careless, if not slipshod. Yet this series is likely to have started in the third century, as the small brooch from Knowth (Fig. 73:2) testifies. Decoration is based on the stylised palmette, and there is some similarity between the decoration here and that of the proto-handpins of Fig. 68.

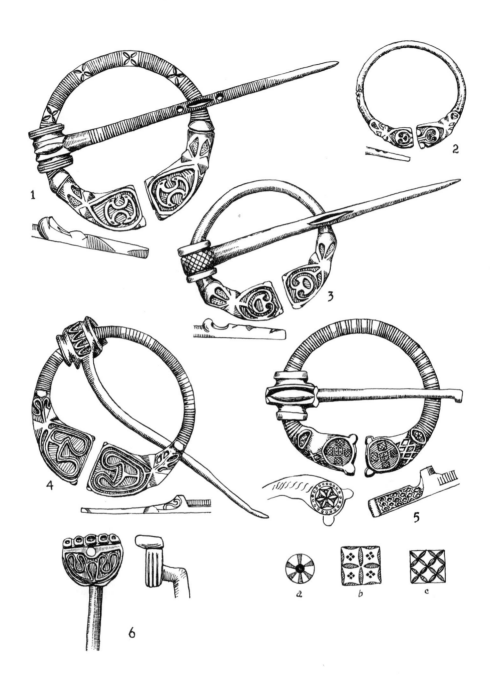

Fig. 73: *Irish Enamelled Zoomorphic Penannular Brooches Group C₃*: 1. Old Castle of Carbury, Co. Kildare. *Group C₄*: 2. Knowth, Co. Meath. 3. Ireland. *Group C₅*: 4. Navan Rath, Co. Armagh. *Group D*: 5. Ireland. 6. Five-fingered Handpin, Trillick, Co. Tyrone. (3/4)

As far as the brooches are concerned, there is overdependence on the stylised palmette as the main decorative motif, and it appears in this and in other badly emasculated versions on the majority of the brooches in this series. One later developed brooch bears a Greek cross in the place of these simple symbols, which must indicate the arrival of Christianity and the spread of Christian fervour. The distribution is mainly Meath-Westmeath, and the workshop may have been remote, since the metalworkers appear to have been unaware of any other forms of motif. Yet a lead pattern for a brooch in this series was found at Dinas Powys, Glamorgan.[18]

Brooches gathered together within the C$_5$ sub-group are clearly the work of one individual who was excessively short-sighted. His work is characterised by its inconsistencies; decoration is often meticulously carried out, the toolwork is sometimes of extreme delicacy, yet no two terminals on the same brooch bear identical decoration. Where lines run concentrically, or are otherwise parallel to one another, definition is so close and so consistent that they remain always equidistant (Fig. 73:4). One must admire so steady a hand. This man was familiar with spirals and other patterns, but he preferred to interpret them in his own individual way. There seems to be little doubt but that he worked in the brooch factory at Clogher, Co. Tyrone, where debris resulting from the production of zoomorphic penannular brooches has been found.[19] The main production period appears to have been in the fifth and sixth centuries; but the C$_5$ brooches may belong to the period when Clogher became the capital of the Airgialla in the fourth century, a date which suits the style of these brooches. Possibly this is the nearest to an exact dating for Irish brooches that it will be possible to achieve. At Clogher there was a complete break with tradition, though what brought it about is not clear: for the brooches which follow the C$_5$ series here are quite different. They belong to another group, Group D, in which the eyes are omitted (Fig. 73:5), and more emphasis is given to the ears, until some assume 'Micky Mouse' proportions. However, these Group D brooches must belong to the fifth century, judged by their association with B amphorae, and they represent the end of the line in zoomorphic penannular brooch development. Rapid devolvement of the Group D style perhaps indicates that manufacture here outlasted the collapse of the local heroic society, following upon the Roman Withdrawal from Britain. The heroic centres, which includes nearby Emain Macha, were sacked in *c*. 450: but the occupational history of Clogher can be traced up till 800. With raw rejects and other signs of brooch-making, the manufacture here of Group D is beyond doubt.

The advent of Group D brooches at Clogher shows that by the fifth century the traditional Group C style, expressed by the short-sighted metalworker, was already moribund. Not a single Group D brooch bears decoration that is other than of millefiori enamel, which was to be expected, since by the fifth century millefiori had superseded all other forms of decoration. The excavations at Garranes[20] clearly showed the extensive use of millefiori at that time. Some particularly fine examples are known. Group D brooch Fig. 73:5 from Ireland has terminals decorated entirely in this style, and each individual square of millefiori has its own particular pattern. Yet it is the handpins which exhibit the widest range of patterns. The three- and two-fingered handpins shown in Fig. 74 demonstrate from their decoration just how expert pattern-making in millefiori could become, the colours used ranging from white to blue, dark red, brown, yellow, honey colour and green — a range of colours wide enough to delight the fastidious. The patterns themselves are exceedingly small, and this small size is achieved by stretching the glass when still hot.

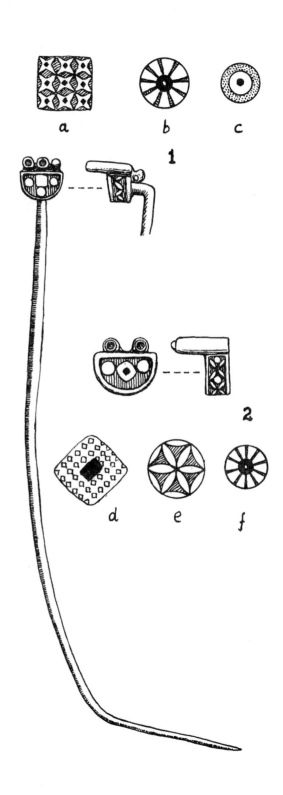

MILLEFIORI ENAMEL

a Red diamonds, white and blue.

b Red centre, white and blue.

c Red centre, white and brown.

d Red centre, honey coloured diamonds on yellow background.

e Blue on white background.

f Red centre, white and blue.

Fig. 74: Millefiori Enamelled Handpins 1. Bartrauve, (Mayo White Sands), Co. Mayo. 2. Ireland. (1/1)

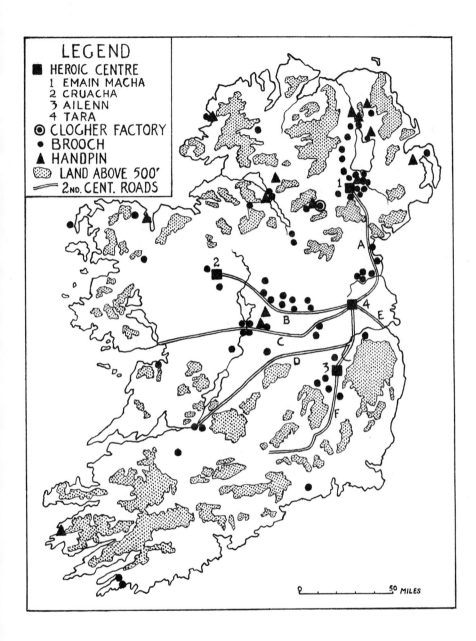

Fig. 75: Map showing Distribution of Zoomorphic Penannular Brooches and of Handpins. Also the situation of the heroic centres and of the 2nd century roads. (A) Slighe Miodhluachra; (B) Slighe Asail; (C) Slighe Mor; (D) Slighe Dala; (E) Slighe Cualann; (F) Road of the Great Wood.

Of all Irish-made bronze objects, zoomorphic penannular brooches are perhaps the most numerous: altogether, there are more than 160 specimens. Some interesting points emerge from the pattern of their distribution. Distribution of the Initial Form was by no means localised; but once brooches started to increase in overall size, immediately the distributive pattern changed. The new area of distribution is bounded on the north by Co. Antrim; on the west by the Shannon; and on the south by the counties Kildare and Laois (map, Fig. 75). An imaginary peripheral line can be drawn around this distribution area, and it will be seen to enclose the main heroic centres of Emain Macha (Navan Fort), Cruacha (Rathcroghan), and Ailenn (Knockawlin), whilst Tara is situated somewhere near the middle. Emain Macha was the seat of the kings of Ulster; Cruacha was the seat of the kings of Connacht; Ailenn was one of the royal residences of the kings of Leinster; whilst Tara (teamair na Riogh) was the seat of the high kings of Ireland, its period of greatest glory extending from the time of Cormac MacArt in the third century until the reign of Dermot MacCearbhaill in the middle of the sixth century. At Emain Macha the kings of Ulster gathered round them a celebrated body called the Red Branch Knights, of whom Cuchulainn was one of the foremost. They were famed for their deeds of prowess and chivalry.

Kings and knights they may have been, but they were also kingly freebooters. Theirs was a life of fighting, feasting and raiding. This was the heroic age in Ireland, and Emain Macha, Cruacha and Ailenn and Tara were its main centres. All these centres were inter-connected by a road system, and the brooch distribution must be studied with reference to these. These centres were riding high on the booty obtained from raids on Britain; but once the source of booty was exhausted, the parasites who lived off it were doomed. Disintegration of their society soon followed. Emain Macha was burned and pillaged when Fergus was king of Ulster — as some say, in the fourth century, though this date is now discounted, and an alternative tradition places this event as late as AD 450.[21] The Roman Withdrawal took place in 410, and the Irish were driven out of North Wales by Cunedda in 440.[22] Apart from the events themselves, Ireland must have been full of refugees, which was too much for the Irish economy to bear, so it collapsed.

The brooch distribution has made it clear that these articles were being made for and bought by the people of these heroic centres. For, after all, fine figures need fine plumage. The overloading of some brooches with decoration often borders on the vulgar. The best made and the most ornate belong to the fourth and fifth centuries, during the period of the decline in Roman power in Britain. But the bonanza came to an end because of the Roman Withdrawal, and this was followed by the flood of refugees from North Wales. Nothing illustrates better the retraction of wealth than do the Group D brooches, which, from being quite ornate go into rapid decline, into plainness, and a reduction in size and in quality. The Clogher factory alone appears to have soldiered on for a while. The story of the handpins is less complete, but their distribution is more northern, and the largest concentration is around Lough Neagh. This was also the home of the C_1 brooches, and the source of both may have been the same workshop.

The fact that Clogher alone carried on with the manufacture of Group D zoomorphic penannular brooches may have some relevance in history: for it is known that the Soghain were dispossessed of their midland territories by the Sept Ui Maini in the fifth century. Perhaps this happening together with the sacking of the heroic centres may be looked upon as a single event. Whether or not these happenings are relevant to the history of metalworking in Ireland is uncertain, but clearly they would affect demand for whatever products there were about; and in that sense Irish metalworking must have suffered a serious reverse.

NOTES

1. H.C. Lawlor, 'Objects of Archaeological Interest in the Lough Neagh and River Bann Drainage Scheme', *Journ. Roy. Soc. Antiq. Ireland*, LXII (1932), p. 209.

2. Kelvingrove Museum, Glasgow.

3. Wallis Budge, *An Account of the Roman Antiquities preserved in the Museum at Chesters*, no. 2226.

4. *V.C.H. Berks.*, II, 247.

5. T.G.F. Patterson and O. Davies, 'The Craig Collection in Armagh Museum', *Ulster Journ. Arch.*[3], 3 (1940), p. 70.

6. J. Raftery, 'A Bronze Zoomorphic Brooch, and other Objects from Toomullin, Co. Clare', *Journ. Roy. Soc. Antiq. Ireland*, LXX (1941), p. 59.

7. Illustrated here by courtesy of George Eogan.

8. See footnote 39, Ch. 4.

9. Fig. 45:6, 7 and 8.

10. E. Rynne and A.B. O'Riordain, 'Settlement in the Sandhills at Dooey, Co. Donegal', p. 63, Fig. 8.

11. M. Richards, 'Irish Settlements in S.W. Wales', *Journ. Roy. Soc. Antiq. Ireland*, XC (1960), p. 133.

12. W. Wilde, *Cat. of the Antiquities in the Museum of the Royal Irish Academy*, p. 559, Fig. 455.

13. C. Johns, 'A Roman Silver Pin from Oldcroft, Glos.', *Antiq. Journ.*, LIV (1974), p. 295, Fig. 6.

14. R.B.K. Stevenson and J. Emery, 'The Gaulcross Hoard of Pictish Silver', *Proc. Soc. Antiq. Scot.*, XCVII (1966), p. 206, pl. XI:1.

15. Macgregor, *Early Celtic Art in North Britain*, I, p. 189.

16. Stevenson[14], p. 209.

17. Leeds, *Celtic Ornament*, p. 137.

18. L. Alcock, *Dinas Powys, an Iron Age, Dark Age and early Mediaeval Settlement in Glamorgan* (University of Wales Press, 1963).

19. See summaries by R.B. Warner, *Excavations 1973 and 1974* (Summary accounts of archaeological work in Ireland).

20. S.P. O'Riordain, 'The Excavation of a Large Ring Fort at Garranes, Co. Cork', *Proc. Roy. Irish Acad.*, XLVII (1942), pp. 77 ff.

21. F.J. Byrne, *Irish Kings and High Kings* (Batsford, 1973), p. 50.

22. See the argument in Alcock, *Arthur's Britain*, pp. 125–8, in relation to this date.

PART FOUR: POST-OCCUPATION STYLES

24. Introduction

As Ireland's heroic society was collapsing, the Christians moved in. The monastery of Nendrum was founded around 445, by whoever was the leader of the Whithorn mission at that time. This mission was ruled, from the middle of the fifth century until its close, by St Machaoi, who is now considered to have been also the founder of the Nendrum monastery.[1] Although Ireland now moved towards a dynastic polity, the emphasis is less on political history and more on ecclesiastical history, since the new religion was providing a stabilising influence. On the other hand, and in contrast, Britain remained in a state of turmoil, after the rescript of Honorius to the tribal communities, instructing them to take measures for their own protection. The separation from Rome had come not as a withdrawal of the legions, for most of them were already in Gaul or Italy, but as a hiatus in the apparatus of government. So the middle of the fifth century brought no rapid or general collapse. Perhaps authority had fallen into the hands of provincial or diocesan councils. South of the Wall sites remained in occupation well into the century. But the year 410 had also seen a renewal of Saxon raids, and resistance to these must have come from the people themselves. Some tribes are known to have started to hit back. In the middle of the century, Cunedda, at the head of the Votadini from Manau (the district centred around Clackmannan) marched south to north Wales in order to drive out the Irish settlers. This is clearly one of the most positive acts of the period, and it probably, indirectly, helped to bring about the collapse of Ireland's heroic society, as we have already seen.

But what brought about the subsequent revival of Celtic art? The normal view of this revival is one of a sudden outburst of activity once the Romans were out of the way. It now seems that the process was more gradual, like the Roman Withdrawal itself. The survival and revival for a time of the old Celtic religion may have been a contributing factor to this revival of the art. Romano-Celtic temples lay out in the countryside, away from the principle centres of population. One was built in the late fourth century within the ramparts of Maiden Castle, after the fortress had lain deserted for centuries. This was a temple of the normal 'Romano-Celtic' type, with square *cella* set within a veranda, the walls decorated with painted panels in blue-green, dark red, and white on a background predominantly terracotta in colour. The latest coin found here was one of Gratian (367–83).[2] Another temple was built within Chanctonbury Rings, in Sussex. At Lydney, Gloucestershire, an elaborate temple was built, again within an earlier earthwork. Perhaps by placing these temples within earlier native fortifications, a persistent tradition was somehow acknowledged. The Lydney temple, dedicated to the Irish god Nodens, a god of hunting and healing, was built at precisely the same period as was the temple at Maiden Castle,[3] so that the rash of temple building at the end of the fourth century looks like a concerted effort on the part of a disenchanted people. The Lydney temple was of unusual design, with its nave and surrounding ambulatory, and several side chapels, all suggesting that it was meant to house a congregation intending to take part in some communal act of worship. The choice of god was equally unusual, since his affinities lay westward with Ireland. The worship of Nodens was confined to Lydney, where it was established not many years before the birth of St Patrick. It is evident that, in the more remote areas of the west and north-west, fourth-century Christianity, for all its official prestige, was no more than first among equals.

Christianity had made headway in the towns and amongst the aristocracy; but the peasantry were still largely pagan. It is unlikely that there would have been much political change for some years to come. Amongst the events of the times was a visit paid to Britain by Germanus, bishop of Auxerre, who had been sent by the Pope to arrest the spread of Pelagianism amongst the aristocratic minority. Germanus, who had once been a general in the Roman army, also took the opportunity to campaign against the Saxons and their allies the Picts. He began his campaign by baptising the British forces, and he taught them to shout 'Alleluia' at the moment of attack. The terror of this sound so unnerved the Picts and Saxons that they fled! Almost the last fixed date in fifth-century chronology is provided by the unsuccessful appeal to Aetius, which is recorded by Gildas. Known as the 'Groans of the Britons', the complaint was that the continental invaders were driving the British to the sea. It is clear that civilisation in the fifth century was running down, and with increasing speed. But if conditions were as bad as we are led to believe, who was able to afford expensive items like hanging-bowls?

The disappearance of mass-produced pottery and its substitution by home-made wares is another sign of the times. This home-made ware bears a remarkable resemblance to Belgic pottery of pre-Roman times. This return to pre-Occupation forms is symptomatic of a bankruptcy of ideas. But if potters had to return to pre-Occupation times for their forms, what prevented the metalworkers from doing the same? Instead, as we look around, all we see are decorative forms which were current in Ireland in the fourth and fifth centuries. Also there are some horrible and clumsy imitations as well. The contrast between the two styles is remarkable. It is clear that there must have been numbers of unemployed metalworkers, both British and Irish, in the fifth century. With the collapse of civil government in both countries, it was only natural that these craftsmen should turn to the one stable body left, the Church. Some metalworkers must have made contact with the Saxons, who came, in the first place, at the invitation of the British; for the earliest permanent Saxon settlements in Britain resulted not from raids or seizure by land-hungry settlers, but by Vortigern's invitation. For Vortigern had turned to the Saxons in his effort to keep out the Picts. It is known that British potters were working for the Saxons. Some of the hand-made pottery from Anglo-Saxon cemeteries has Saxon decoration imprinted on a traditional fabric, clearly indicating that native potters were working for new masters.

Nevertheless, it is probably still true that the hanging-bowls found in Saxon graves were booty. If booty, then the original owners must have been ecclesiastical. Metalworkers' workshops existed in Whitby Abbey. One hanging-bowl has escutcheons decorated with a cross supported by two dolphins — symbolising Christ. If booty, then these objects were stolen from Christian churches, where some say they were used as sanctuary lamps, whilst others maintain that they were votive offerings. To a pagan Saxon a votive chalice would have been mystical: perhaps he regarded it as having magic properties. Either that, or he was merely acquisitive.

With the passage of time, authority passed to the Saxons. The anonymous Gallic Chronicle of 452 records that 'Britain, long troubled by various happenings and disasters, passed under the authority of the Saxons'. This is not strictly true, since a long period of fluctuating warfare culminated at some date before 500 in a British victory at Mount Badon, after which there was peace for two generations. But the Saxons came back in 550, and by 600 most of Britain had fallen to the Saxon kingdoms, with the exceptions of the Cornish peninsula, Wales, parts of the Pennines and the north-west, and Scotland.

It is difficult to equate artistic development with these events. The hanging-bowls bear escutcheons and prints both having scroll decoration on an enamel background. These are the chief expression of the revival. They range in time from the late fourth to the seventh centuries, so that a long stylistic development is implied. This development was vigorous at first, and apparently was pursued without interruption. At the same time we do not know how vigorously the Christian church was pursuing its proselyting activities in Britain. The Irish Sea had become a Celtic Province, whilst in Ireland Christianity was steadily making inroads. Perhaps the Christian west met the pagan east midway along the Fosse Way: but with Celtic-speaking Britons in the Fens, where they were still living as late as 700, conversions among the Saxons may have been a daily occurrence; and converts would be just those people who would be most likely to present a votive chalice to the church, either in thankfulness or as a guarantee of luck.

The revival and the flowering of the art has to be seen against both a fluid situation and the establishment of the Saxon kingdom. But, who revived it? In the distribution of hanging-bowls there is nothing that would provide a clear answer. The bowls are so expertly made that they must be the products of workshops where the staff was both expert and highly trained; and this remark brings us back to the unemployed Irish metalworkers, who, at this period, would have been the only ones left with the necessary experience. This idea is not new, for others have tentatively put forward the same idea, notably Mlle Henry. But now we go a step further, for it seems that hanging-bowls resulted from two traditions — one Celtic, the other Romano-British. This was hinted at by Kendrick more than 40 years ago. This second tradition adds complexity to the situation, which must be seen against the fluid situation of the times and the increasing inroads being made by Christianity, much of which was under the authority of Irish missionaries.

NOTES

1. E.S. Towill, 'St. Mochaoi and Nendrum', *Ulster Arch. Journ.*, 27 (1964), p. 103.

2. Wheeler, 'Maiden Castle', p. 131.
3. Wheeler and Wheeler, 'Lydney Report', p. 23.

25. Craftsmanship in Ireland (III)

As we have seen, the sacking of the heroic centres, coupled with the dispossession of Soghain territories, both events of mid-fifth-century occurrence, undoubtedly left many craftsmen with little or nothing to do. As a result of these twin disasters, it is not to be wondered at that objects decorated in the Soghain style, as exemplified by the decoration on the B_2 sub-group brooches, are depressingly few, in what, for lack of a better term, might be called the post-heroic period. But we are permitted a glimpse or two of what must have been about in those days.

The only worthwhile objects are those illustrated in Fig. 76, though one of these is outside our period altogether. This is the hanging-bowl escutcheon (Fig. 76:6) which bears decoration in the chip-carving technique. This small collection of objects may not be very impressive; but the standard of workmanship remains high, and one object, Fig. 76:3, has decoration superbly done on a very small area. Workmanship of this quality can only be the result of repetition; so where are the others? Because of the rarity of decorated items, the period from the mid fifth to the seventh century is dark indeed, and it is only illumined by the sixth century records. In one way, the Christians were to blame, for Christians were not buried with their worldly goods; whereas in Britain, hanging-bowls were being carefully preserved in Saxon graves.

So, although the present collection is small, it does provide some useful snippets of information. Firstly, it is noted that all spiral work is triskele based. Of all the former motifs in Celtic art, the triskele and the spiral are the lone survivors. All Irish triskeles of the period have arms that curve to the right. This was true of the Bann disc (Fig. 35:2), the Lambay scabbard mount (Fig. 41:1) and the latchet illustrated in Fig. 67:2. A vacant space is always left at the centres of these triskeles, though occasionally it is occupied by a circle, as here. Triskeles of this precise form are not found amongst Celtic decoration in Britain.[1] If there are similarities, arms invariably curve to the left. This is a simple rule-of-thumb method for distinguishing Irish from British triskeles. This rule will also help to confirm what is well known, and that is that the Tal-y-llyn triskele (Fig. 22:3) was under strong Galloway-style influence.

The best item in the collection is Fig. 76:3, which is a round plaque with a central decorated boss, and it comes from Lagore, Co. Meath.[2] Unfortunately, it was unstratified when found; and since there is no information about the beginning history of this crannog, the matter of date must be set aside for the moment. The decorative style is nearest to that of the early hanging-bowl escutcheons; and the method of extending the triskele arms to form triple spirals is the kind of work one is used to in Britain. Note how the spiral finials swell out to form trumpet-like ends, which again is a characteristic of the escutcheon spirals. The manner in which these trumpet-like finials are disposed occasions

Fig. 76: Craftsmanship in Ireland 1. Enamelled Bronze Button, Garranes, Co. Cork. 2. Decorated Disc, Ireland. 3. Enamelled Disc, Lagore, Co. Meath. 4. Enamelled Toilet Implement, Stoneyford, Co. Kilkenny. 5. Punched Design, Moylarg Crannog, Co. Antrim. 6. Hanging-bowl Escutcheon, River Bann. 7. Decorated Brooch, Nendrum, Co. Down. 2, 3, 4, 5, 7 (3/4); 1 and 6 (3/2)

some interesting observations. There is a total number of three: one is utilised to form the one arm of the triskele, whilst the remaining two are linked to their opposite numbers on adjoining spirals. It is the manner in which this linkage takes place and its form that gives a clue to the origin of these trumpet-like finials. Pertinent to their development must be the scrolls linking running spirals, mostly single, but some double, on the enamelled disc,[3] unfortunately without provenance, shown in Fig. 76:2. Clearly, these are based on the broken-back scroll, for examples of which one should turn to Figs. 34 and 38. On the other hand, it would have been possible for the earlier trumpet forms to have survived long enough to have influenced the shape of these spiral finials; and one immediately thinks of the sinewy and elongated spiral forms on the Bann disc. But the former explanation offered would seem to be the more probable. Whilst broken-back scrolls are not of common occurrence in Britain, they are essentially an Irish development, in which the Loughcrew bronzesmiths showed a strong hand.

This unlocalised disc (Fig. 76:2) is not representative of Irish craftsmanship at its best; for the running spirals are not well executed, and not all of them are singles, since others are poorly executed doubles. But all the single spirals resemble those on the Dooey trial piece, and they could be representative of a current style. The decoration within the central roundel has been badly conceived, whilst its meaning and purpose remains obscure. It repels rather than attracts the eye. But the style seems to be early, so that one might subscribe to a fifth-century date for it. To this period of uncertainty must belong the toilet implement of Fig. 76:4, which comes from Stoneyford, Co. Kilkenny.[4] Though the basic form of the implement is classical, the enamel-backed design on the roundel proclaims its Celtic origin. The overall pattern is a triskele with the ends of its arms extended to form triple spirals, but the workmanship is not good. The presence of petals in the panels both above and below this roundel of decoration proclaims that the piece is of fourth/fifth-century workmanship. But an object nearer to AD 500 must be the decorated enamelled button, Fig. 76:1, which was found in the multivallate fort of Garannes, Co. Cork.[5] Here the triskele form is more firmly represented, and, considering the small size of the object, it is well executed.

There is no information here that would make it possible to date the Lagore disc (Fig. 76:3). The excavator was unable to date the first occupation of the crannog,[6] even though finds included sherds of figured Samian. But since they came from disturbed ground, their importance is minimised. The disc was also unstratified. That there must have been many more curvilinear designs like this about at that time is suggested by the punched design (Fig. 76:5) on the base of the strainer from Moylarg crannog, Co. Antrim.[7] It compares very well with the spiral pattern on the Lagore piece. The strainer is in the form of a skillet, a circular and round-based vessel of sheet bronze, with a narrow flat out-turned rim. It was provided with a handle. Clearly, vessels such as these were made in imitation of Roman paterae; and long ago Armstrong[8] illustrated several, all of which were found in Ireland. But whereas the Roman paterae possess handles which are pierced for suspension, the Irish handles are plain, and a further distinction relates to the rounded bases. Roman paterae have flat bases: they were largely current in the second century; but these Irish skillets are unlikely to have been so early, perhaps dating from at least two centuries later, or maybe a little later. Two skillets of this type were found in the Saxon monastery of Whitby.[9] The foundation date of the monastery, 657, provides a *terminus post quem* dating, showing that skillets in this form were still very much in use in the seventh century. It is more than probable that these Whitby skillets share a common origin with the Irish examples, both lots being distinguished by lack of effective strengthening of handle, rim and base.

A hanging-bowl escutcheon, Fig. 76:6, was dredged up from the bed of the River Bann. Perhaps it should not be included here, since it must be dated by the Hiberno-Norse decoration which it bears, and carried out in an inferior *Kerbschnitt* style. But running spirals decorate the circumferential area, though the quality of workmanship is not good. The persistence of the hippocampic head is interesting: derived from Brigantian sources, in Ireland this form of head was first noted on the Petrie Crown. Above the collar on this escutcheon are spherical triangles with central dots, and all in juxtaposition, again survivors from a previous age. Spherical triangles in this form are the equivalent of an Irish trade mark. Yet the basic design of the escutcheon, with its central diamond-shaped void and two flanking palmette-shaped voids is British, several of this precise form having been found from Warwick to Moidart and Inverness. In Britain the type is either plain or decorated, the plain specimens being of northern provenance. These escutcheons are discussed in the following chapter. Lastly, from Nendrum, Co. Down,[10] comes the small penannular brooch, ornamented with double spirals that are formed of two swollen finials conjoined.

This Irish contribution to late and post-heroic period metalwork is seen to be of more than passing interest, and if its showing is limited, this is not due to a discontinuance of metalworking, but to the paucity of finds. The small numbers are due in part to a lack of systematic excavation work, as Richard Warner's excavations at Clogher are at present demonstrating. The only known pseudo-zoomorphic brooches in Ireland have come from excavations at Knowth, New Grange and Tara: and until more digging is done on domestic and fortified sites of the period in question, the Irish story must remain incomplete.

NOTES

1. Except where Irish influence can be proved.
2. H. O'N. Hencken, 'Lagore Crannog', *Proc. Roy. Irish Acad.*, LII C (1950), Fig. 11:310, and pl. XIV:1.
3. A. Mahr, *Christian Art in Ancient Ireland* (Hacker Art Books, 1977), I, pl. 1:1.
4. Ibid., pl. 41:5. Also S.P. O'Riordain, 'Roman Material in Ireland', *Proc. Roy. Irish Acad.*, LI C (1947), p. 64, Fig. 6:4; J.D. Bateson, 'Roman Material in Ireland: a re-consideration', *Proc. Roy. Irish Acad.*, 73 C (1973), p. 73.
5. S.P. O'Riordain, 'The Excavation of a large earthern Ring Fort at Garranes, Co. Cork', *Proc. Roy.*

Irish Acad., XLVII C (1942), Fig. 3:231, pl. XXIII:1.
6. Hencken[2], pp. 1 ff.
7. *Journ. Roy. Soc. Antiq. Ireland*, 24 (1894), p. 317.
8. E.C.R. Armstrong, 'The la Tène period in Ireland', pp. 1 ff.
9. C. Peers and C.A. Ralegh Radford, 'The Saxon Monastery of Whitby', *Arch.*, LXXXIX (1943), p. 66, Fig. 17:1 and 2.
10. *Arch. Survey of Co. Down*, Fig. 80.

26. Hanging-bowls

In terms of skill and artistry, few products are the equal of hanging-bowls. They are of hemispherical form, and more often than not, they have concave bases. They were made of beaten or spun bronze, having thin walls which are shouldered at the top. The metal either thickens at the rim, or it is bent over to make a flat top. Suspension was presumably by means of chains (none has yet been found) which were attached to hooks springing from kite-shaped or circular escutcheons sweated on the bowl just below the shoulder. Escutcheons were either three or four in number: most are ornamented with elaborate spiral patterns against a background of enamel. Normally, both inside and outside there are circular decorated prints on the bases of these vessels. Hanging-bowls are not very large, varying from as little as six inches to roughly one foot in diameter. Although of native manufacture, most of these vessels have been found in Saxon graves and on domestic locations.

Because nothing definite is known about their purpose, hanging-bowls have provided a fruitful field of investigation for many amateur detectives. Most writers regard them as being sanctuary lamps; and recently K.R. Fennell[1] has put in a plea in this connection. Mlle Henry said the same thing many years ago.[2] The argument gained strength from the fact that one vessel possesses escutcheons engraved with a cross and two fishes or dolphins, symbols of Christ. At the monastery of Whitby, founded 657, several fragments of hanging-bowls were found, so the excavators[3] assumed that these vessels had been utilised as lamps, and they gave references to back up this contention. Another theory is that these vessels were pressed into service as liturgical water vessels. The basis of that thought is Bede's story of King Edwin's authorisation for them to be used for the purpose of providing drinking water to travellers. But Celtic churches were influenced a good deal by continental ecclesiastical customs, and on the Continent it was customary to present votive chalices to be hung up inside the churches. For this reason, Monsignor McRoberts[4] looks upon hanging-bowls as being votive gifts, given to Celtic churches for the enrichment of their interiors. Since the association of some of these vessels with religious foundations is not in doubt, the theory gains acceptance. By any standard, hanging-bowls must have been costly objects, and their elaboration both inside and outside with enamelled prints and escutcheons renders them impractical for ordinary daily service. Therefore, they may be regarded as expressions of piety and generosity on the part of the donor. In present day Greece brass censers are used for the same purpose.[5] These censers are suspended from the roof of the church; and once all available space is taken up, a new church, alongside the first, must be built, and the whole process begins all over again.

This is not to say that hanging-bowls were used exclusively for this purpose; for hemispherical hanging-bowls date from la Tène times. In Britain, hanging-bowls have a long

236

history, starting with the specimen from Cerrig-y-drudion.[6] However, the type with which we are concerned is more likely to have originated in Britain during the Occupation. Kendrick[7] was the first to see an association between bowls and some Roman objects; and from this association he concluded that the whole series was the work of British craftsmen, whose initial efforts he placed in the period AD 300–400. It will be recalled that this was the period which saw the beginnings of the expansion of the old Celtic religious houses, evidenced by the building of new temples on former and mostly disused native sites. Christianity itself had become the religion of officials, so that one or other of these two religions could have been responsible for creating a demand for votive offerings in these troublous times. There has got to be some reason for the emergence of the hanging-bowl, as we know it, at this time.

The bowls, being of very thin metal, do not always survive, and those that have rotted away are often represented only by their escutcheons and prints. The earliest escutcheons appear to be those from the Roman fort at Newstead.[8] There are four escutcheons, and they are long, plain, and kite-shaped; they are hand-wrought and hollow, with very convex surfaces. At the top of each escutcheon there is a primitive-looking loop, bent *outwards* over a free-moving ring, held captive. Since Newstead was evacuated *c.* AD 180, these escutcheons must be amongst the earliest in existence in Britain, though others, of a somewhat similar form, were found in the hoard of Roman silver discovered at Traprain Law.[9] Here loops have given way to hooks with ornithomorphic terminations, so that a certain amount of development had taken place by the fourth century, which is when the hoard was hidden. Bowls were also coming into the country from the Continent. One escutcheon, from Barton, Cambridgeshire,[10] is exactly similar to others on the bowl from Nauheim,[11] whilst a heater-shaped specimen from Silchester[12] has an exact parallel at Sackrau, Silesia. These continental hooks are bent *inwards*, in contradistinction to the Newstead and Traprain Law specimens. The bent-inwards hooks are more suited to keeping the rings captive, since they face the wall of the vessel; and once the escutcheon has been sweated in position, the ring remains captive for the life of the vessel. Another point of interest is that, in the case of the continental bowls, there are three escutcheons; whereas at Newstead there were four. This comparison may or may not be important, since among the British specimens, some bowls have three escutcheons whilst others have four. Continental bowls also tend to be squat.

Kendrick thought that the British bowls owed their hemispherical form to what he termed the Irchester type. The Irchester type is a hemispherical bowl of beaten bronze, having a cupped base and a short inturned neck and rim, and the form is of fourth-century date. However, another candidate in the prototype stakes is the hanging-bowl from Finningley, Yorkshire,[13] which meets most of the requirements in that the rim and neck are in the Irchester tradition, but that there are three escutcheons having *inturned* hooks with hippocampic features. This bowl could conceivably be earlier than the Irchester type, an opinion based on the similarity that exists between the knobbed 'manes' of the Finningley hippocamps and others that are found on a rather rare type of brooch, examples of which have come from Caerleon,[14] Richborough,[15] Corbridge[16] and also from Heddernheim.[17] These British brooches appear to be of third-century date. This may be too early a period for the Finningley bowl, but the similarity amongst all these specimens is too striking to be ignored.

Should the Finningley bowl be accepted as one of the earliest in the British series, then the next noticeable point about it is that it bears no Celtic ornamentation at all. This absence of what might be called true Celtic ornamentation (perhaps 'native' would be a better qualification) has been noted on some bowls, on their escutcheons and prints, and immediately it raises a query. Considering what happened in the late first and second century, when alien designs and patterns found their way into people's preferences, it might be thought to have been the result of the art going underground. But, in contrast to these, there are other escutcheons and prints decorated with designs that are wholly Celtic in character. So that here there is the situation in which, on the one hand, there are prints and escutcheons decorated with designs which are definitely not Celtic, or, alternatively, with designs which appear to be poor copies of what Celtic designs should be; and on the other hand there are prints and escutcheons decorated with a particularly pure form of Celtic art. The differences between the two styles are so wide as to suggest two differing origins. This situation came about gradually, not suddenly; and with it went the gradual realisation that the art was back, in all its former vigour, with some magnificent scroll work for everyone's delectation. Where did it come from, this art? And why should its representation on one set of bowls vary from another? Perhaps part of the answer is that some bowls differ from the rest. Some have rims that remain vertical, but thicken gradually above the shoulder, clearly to give strength to the rim. Others have a wide and flattened lip resulting from a fold in the metal, by its having been pressed inwards and downwards until finally it ended up in a horizontal position, giving even greater rigidity to the rim. Next we note that escutcheons and prints with decoration that is only remotely Celtic, or not Celtic at all, all belong to the first form of rim; whereas escutcheons and prints bearing top quality Celtic ornamentation all belong to bowls with the second type of rim. Obviously, here we have two distinct groups, which differ widely, one from the other. To the first group we will give the letter 'A', and to the second group the letter 'B'; and if there is any relationship at all between these two groups, it can be said to have been on the same basis as Group B was with Group C in the matter of the zoomorphic penannular brooches. This fact has already been noted by Mlle Henry.[18]

Next, there is the matter of the association of escutcheon types. In the case of Group A bowls, escutcheons are either plain heater shaped; slightly ovoid with advanced palmette-shaped voids; or, if they are circular, they bear bastard Celtic ornament crudely executed, or non-Celtic patterns. The workmanship is careless, often crude. In the case of Group B bowls, escutcheons are usually circular, normally are expertly ornamented with genuine Celtic designs in the form of triskele-based triple spirals with ends linked by trumpet scrolls, all against an enamel background. Group B bowls appear in the fifth century in fully developed splendour; but there are earlier beginnings. Everything points to a long tradition in the making of patterns, of which there are few indications in Britain. Fully fledged patterns such as these did not evolve overnight. So the situation we are examining is likely to turn out complex, and for this reason many conflicting statements have been made. Even though evidence for the development of these spirals is lacking in Britain, yet Leeds suggested that this art was taken from Saxon England to the Celtic west.[19] Statements along these lines continue to be made. Yet in Ireland there was a continuing tradition of spiral making, on brooches (Fig. 70); on latchets (Fig. 67); and on the objects shown in Fig. 76.

Division of the bowls by rim form into two recognisable groups both simplifies and clarifies the history and development of British hanging-bowls. Group A bowls are Romano-British, as Kendrick would have wished: their history and development is fairly continuous from small beginnings in the earlier Occupation period. Group B bowls also show a development, chiefly in the escutcheon form and decoration; but here that development appears to be aloof from happenings other than those in the Celtic areas. Insular sophistication seems improbable except by means of help from outside. But Group A must be essentially British, the bowls have the widest distribution in Britain, and even an escutcheon mould has been found in the far north near Inverness. With these most people would include the Wilton bowl, the escutcheons of which have a quadripartite design of four peltae.[20] Roman influence on this pattern is clear. Escutcheons and bowl are bigger and heavier than most; whilst the bowl itself tends to depart from the A Group, without actually doing so. Both Leeds and Kendrick believed that the Wilton bowl belonged to the early fifth century: this, of course, is only a guess, but it is a satisfactory guess, and one that gains acceptance here.

Group A

These Romano-British, or Romanising bowls (as Kendrick preferred) never rise to the level of sophistication that might be expected: the reason being that in most cases there is no standardised form for the escutcheons, amongst which there are three distinct types. Decoration, on the whole, consisted of anything that came to hand. But the bowls themselves are remarkably regular, with thickened rims, short hollow necks, and pronounced shoulders.

We can neglect kite or heater-shaped escutcheons, and get straight down to a consideration of a very individual form of escutcheon: that slightly ovoid form with central diamond-shaped void, flanked by two other voids in the shapes of stylised palmettes (Fig. 77). A mould for casting this form of escutcheon was found at Craig Phadraig, near Inverness[21] (Fig. 77:2). There are not many escutcheons of this form, but they are widely distributed. The hooks always have zoomorphic features, whilst the escutcheon itself is plain, but for the simple mouldings it bears. Each and every one possesses a collar, above which there is a moulding of some sort. The plain specimens are the escutcheons from Eastwell, Leicestershire;[22] from Castle Tioram, Moidart;[23] and Craig Phadraig. Apart from variations of size, they hardly differ at all. All are collared, with seed-shaped mouldings above, except in the case of the Eastwell specimen, on which these mouldings appear to be higher up the neck. There is a common zoomorphic interest in the hooks, which basically are similar. The hooks on the Wilton bowl, mentioned above, are also similar, though perhaps more stylised. The hippocampic hooks on the Finningley bowl may have influenced the treatment of the hooks here and started a custom.

Three bowls had decorated escutcheons; these are the bowls from Hildersham, Cambridgeshire[24] and Baginton, Warwickshire;[25] whilst from northern Ireland comes a single decorated escutcheon (Fig. 76:6). The Hildersham escutcheon is the largest of the form, and it carries some horrid, jagged decoration that takes from its bold simplicity. One thinks of something Roman, but one would be wrong, since these saw-teeth appear on some pre-Occupation objects, like the Polden Hill mount (Fig. 24:3) or the Trelan Bahow mirror. The maker of this bowl was endeavouring to copy something, and not doing it very well; or he was working from memory, as he was with the print decoration (Fig. 78:4). There is often disharmony between print and escutcheon decoration on some bowls. Here is a case in point. The reason is not clear. But this has been a serious attempt to copy decoration that is peculiar to the Group B bowls; and the same thing was

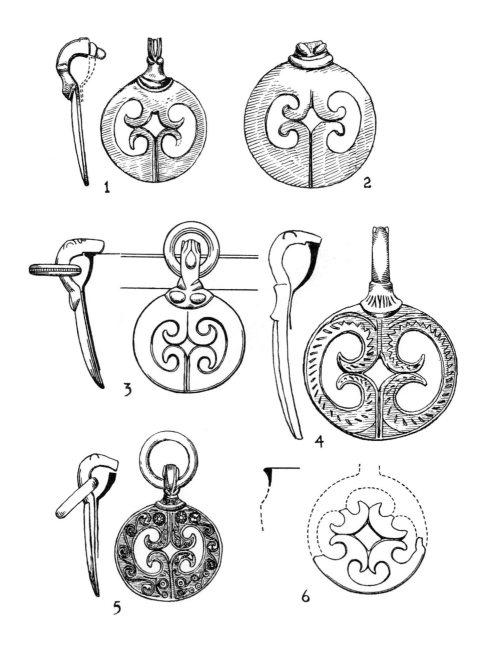

Fig. 77: Ovoid Hanging-bowl Escutcheons with Palmette-shaped Voids, Group A 1. Eastwell, Leicestershire. 2. Impression from Clay Mould, Craig Phadraig, Inverness. 3. Castle Tioram, Moidart. 4. Hildersham, Cambridgeshire. 5. Baginton, Warwickshire. 6. Tummel Bridge, Perthshire. (3/4)

Fig. 78: *Romano-British Triskele Forms* 1. Kingadle, Carmarthen. 2. Verulamium, St Albans. *Bronze Plate with Palmette-shaped Voids* 3. Richborough, Kent. *Bronze Hanging-bowl Enamelled Prints* 4. Hildersham, Cambs. 5. Baginton, Warwickshire. 6. Bekesbourne, Kent. 7. Whitby, Yorkshire. (3/4) except 1 (3/8)

attempted again in the case of the Barrington (Edix Hill Hole) Cambridgeshire print, which, as Mlle Henry has pointed out, is in the same style as the zoomorphic penannular brooch from Ireland (Fig. 70:2). The ornithomorphic finials are again Irish. What can one make of an instance like this, except to say that in both cases, Hildersham and Barrington, there were copyists at work; and what a mess they made of the triskeles.

So far as the Hildersham piece is concerned, the craftsman had lost track of what he was doing when he came to the third arm of the triskele, so he abandoned it, waving goodbye to progress. Actually, the basis of the layout is the simple Celtic pattern shown in Fig. 81:1; but three stylised palmettes have been introduced gratuitously, presumably as a reminder that the escutcheons had palmette-shaped voids. In the case of the Barrington print (Fig. 82:4), the craftsman was aware that triskeles had only three arms; but he had four secondaries with ornithomorphic finials with which to fill up space, so he sneaked in a fourth arm, but he made it tenuous, hoping that nobody would notice that he had to complete his pattern in this way. These two bowls must have been contemporary with some of the Group B bowls, otherwise the man would have had nothing from which he could copy his patterns. This is a point to be remembered.

The escutcheons of the Baginton bowl are also decorated, this time with running scrolls and leaflets, originally filled with red enamel (Fig. 77:5). This time it is the scrolls themselves which are enamel filled, with is contrary to Group B practice, in which enamel forms the background to the designs. This reversal of practice also stamps Group A escutcheons as belonging to a different tradition. The same remarks apply to the print (Fig. 78:5) which is a more ambitious production; but here again, as with the escutcheons, the spirals are enamel filled. Broken-back scrolls are particularly evident, and the nearest to these are those illustrated in Fig. 76:2. The centrally placed hexafoil motif has been executed in its true form, a reversal of Roman practice. The occurrence of a specimen of this ovoid form of escutcheon in northern Ireland is evidence of contact, for it is a local production, and not an export. It is also of much later date.

Up till now, escutcheons have had only two palmette-shaped voids. Now we have an incomplete specimen from Tummel Bridge, Perthshire[26] which has a quadripartite design of four palmette-shaped voids. A similar design, but with a reversal of the palmettes, can be seen on a bronze plate (Fig. 78:3) found at Richborough, Kent,[27] on Site III, and associated with late pottery. But a most interesting variation of the quadripartite palmette design appears on a print (Fig. 83:6) from Chesterton-on-Fossway, Warwickshire,[28] a bowl that clearly belongs to Group B. Here the whole pattern is carried out in interlinked scroll work, resembling spirals without being true spirals; and the incidence is interesting in that contemporaneity could be suggested for escutcheons like Fig. 83:4 and the Tummel Bridge escutcheon. If further confirmation is ever forthcoming, then we are well on the way to showing that (so far) the histories of Group A and Group B were running a parallel course.

This matter of date is most tantalising. For hanging-bowl prints or escutcheons, or the complete bowls are so often chance finds; or where associations occur, as when the bowls are found in Saxon graves, any hope of an absolute date remains unrealised. These bowls neither date themselves nor the objects associated with them. But there are pointers. For instance, the Tummel Bridge escutcheon was found in association with three silver brooches, two being slightly decorated, one with a double line of dots near the edges of the large terminals, the other with dots on the flat pin's widened centre and ridging on the head. For these, Mrs Fowler[29] inclines to a fifth- or even a fourth-century date. A measure of confirmation is provided by the Richborough evidence; for here, it will be remembered, the bronze plate was associated with late pottery – how late we are not told. But perhaps a *terminus ad quem* date of the first years of the fifth century might be suggested by the evidence, which implies that designs such as these did not have to wait for the Roman Withdrawal before they made their appearance. Perhaps Mrs Fowler's inclination to a fourth-century date should gain support.

Fig. 79: Enamelled Print from Large Hanging-bowl, Sutton Hoo, Suffolk. (3/2)

But what of those Group A escutcheons with Romano-British decoration? Admittedly, these pose a lot of problems for which there are few answers. Not always do escutcheons reflect the tendencies of the day if we do not include a reference to the prints: when both occur on the same vessel, they must be studied side by side. This is plain common sense, since the prints provide an additional outlet for artistic expression, and it is curious to observe how often the two differ from one another. A good example of what is meant is provided by the Hildersham escutcheons and print. The escutcheons give no clue to what effect Celtic art was having on the craftsman's mind, but the print does. Another example

Fig. 80: Bronze Hanging-bowl Escutcheons and Prints 1. Escutcheon and Print, Barlaston, Staffordshire. 2. Escutcheon, Faversham, Kent. 3. Escutcheon and Print, Dover, Kent. 4. Escutcheon and Print, Scunthorpe, Lincolnshire. 5. Escutcheon, Northumberland. (3/4)

is the Barlaston, Staffordshire[30] bowl (Fig. 80:1) on which escutcheon decoration is at complete variance with that on the print. The print has a large circular void at its centre: the remaining ring of metal is ornamented with linked palmettes, but their character is Irish, of a form seen on zoomorphic penannular brooches, like Fig. 72:1, or Fig. 73:2 and 3. Leeds[31] recognised this similarity many years ago; and, in addition to the brooches, he mentioned also handpins (Fig. 71:5). What can one say of this Barlaston decoration except that the craftsman responsible for it must have been copying from some Irish metalwork which he had seen. This print decoration caused Kendrick[32] to misjudge the issue, for he included the Barlaston bowl in his 'ultimate la Tène Group', believing it not to be later than *c*. 300. Kendrick was at least a century too early; for a better clue to date is given by the escutcheon which has an insert of millefiori enamel at the centre of the pattern. Millefiori enamel appears not to have been endemic in Britain, but was in all probability introduced into the country from outside. It was commonly used in Ireland on zoomorphic penannular brooches and on handpins, most likely during the fifth century. The so-called swastika design on the Barlaston escutcheon really consists of four triskeles which are conjoined at the centre of the escutcheon and to one another by ill-conceived broken-back scrolls. It seems that the Romano-British mind dealt in even numbers, contrary to the Celtic mind which dealt in odd numbers. The plain heavy triskeles can be compared with one found at Verulamium in a fourth-century stratum (Fig. 78:2), the triskele being one of the motifs that managed to survive the Occupation. An earlier example comes from Kingadle (Fig. 78:1) where it was associated with third-century coins.

So, on the Barlaston escutcheons and print there is decoration taken from at least two different and differing sources, and in traditions which the craftsman did not really understand. This ability to take decoration from any source is characteristic of all Group A craftsmen and their work. This habit illustrated well a bankruptcy of artistic talent; so we are left with a picture of a craftsman searching round for something to put on his prints and escutcheons. The same remarks apply to all the other items shown in Fig. 80. Even the Christian church had something to throw into the kitty. The Faversham, Kent,[33] escutcheon (Fig. 80:2) has four voids, the forms of which are controlled by the shape of the incised cross, which is flanked by two fishes or dolphins, symbols of Christ. The hook has a hippocampic head. This is a well-finished piece and it speaks strongly of some ecclesiastical connection. The Fig. 80:4 escutcheon has a plain equal-armed cross at the centre, and it is surrounded by a circle of rope pattern. A circle of hanging palmettes completes the border decoration. This escutcheon belongs to a bowl from Scunthorpe, Lincolnshire,[34] the print of which is decorated with a horrible simulation of spirals, really linked circles enclosing chips of millefiori — all in running form between the edge and a central roundel containing four of these millefiori-filled circles linked by hollow scrolls: a veritable jumble of all-sorts, with nothing of immediate import. But worse is to come. There are two forms of decoration on the Dover, Kent,[35] escutcheons, one of which is shown here (Fig. 80:3), the decoration on the second escutcheon being similar to that on the print (Fig. 80:3). All are 'Romanising' in the worse sense, and for that reason they may be earlier than the Barlaston specimens. The Dover escutcheon is shown here because there is scroll work upon it, whilst at the centre there is a crude representation of the hexafoil motif. There was always a possibility that the Celts could put a double meaning into this motif. Here it is defined by the enamel filling, which tends to highlight the little triangular islands of metal which are left proud of the enamel. Their significance was not missed by the craftsman, who put a dot at the centre of each one. So now he had six little spherical triangles executed in the Irish manner. There is nothing in British art to make him aware of what he was doing, so once again the inspiration must have been Irish. On the other hand, the horrid foliate patterns, if so they may be called, were borrowed from the Romans. It is now becoming clear that when the art went underground some time after 196, it was buried for good, at least in Britain. But more power to these men who were struggling to revive an old industry, but without the benefit of handed-down experience. Even the escutcheon from far-away Northumberland (Fig. 80:5) in its decoration shows the same lack of experience.

One can search amongst all the Group A bowls and their bits and pieces for a single good representation of at least one motif, but one will not find it. Those poor simulations of Celtic motifs – the triskeles of the Northumberland and Hildersham escutcheons and prints, the 'Irish' running palmettes of the Barlaston print, the miserable running scrolls of the Dover escutcheon – all these must be seen as reflections of other art patterns and motifs, perhaps seen but not possessed, and current at the time, representations of a reactive art movement. It was inevitable, perhaps, that the two traditions – that is if one is allowed to call the 'Romanising' a tradition – should meet somewhere, and this they did at Lovenden Hill, Lincolnshire,[36] where, in the Anglian cemetery, bowls belonging to both Group A and Group B were found. The Group A bowl had escutcheons decorated with coarse linear decoration, consisting of large, limp ill-drawn swasktikas, with their ends twisted into spiral forms. An escutcheon of the Group B bowl is shown in Fig. 83:5. Naturally, both bowls had been purloined, but, though not found together, there must be some degree of contemporaneity, indicating, from the point of view of time, some measure of support for some of the statements made above.

Group B

The hanging-bowls of this group are distinguished by their folded-over rims, a method of construction that gives greater rigidity to the rim than did the method adopted for Group A. Seemingly, this type of rim was adhered to throughout the whole existence of Group B. The bowls are embellished with escutcheons and prints on which the designs are in the true Celtic art tradition, without any deviations. If anything, these bowls have a more pronounced kick at the base, giving them a slightly more squat appearance. Quite clearly the bowls were meant for table use as well as for hanging. This is why there is a print both inside and outside. This dual usage has encouraged the belief that some of these bowls were used as receptacles for communion wine. The association of bronze with the Body and Blood of Christ was an early one, for it was at one time believed that the nails which fixed our Lord to the cross were made of this metal.

This discussion will begin with a look at the large hanging-bowl from Sutton Hoo.[37] This decision may cause surprise: but, for reason, there is the fact that this bowl does not fit into any series, and in that respect is quite unique. But it is included with the Group B bowls because of its decoration, which is pure Celtic. But it is Celtic with a difference, in that the overall pattern of spirals is made up of elements not seen either amongst the art patterns of Group B or indeed on any other works of art up to and including the seventh century. Agreed, there are ornithomorphic finials (Fig. 79), but they are ornithomorphic finials with a difference, in that there are none quite like them. Perhaps the nearest to them are to be found on the zoomorphic penannular brooches from Bloomfield, Co. Roscommon (Fig. 72:3). This spiral decoration displays a clever use of the compass, and they are so finely executed that the toolwork may be said to be equally clever. And all this is on a form of bowl that is nearer to Group A than it is to Group B. Moreover, the central roundel both consists of and encloses many patterned circles of millefiori against a background of red enamel, a fact which encouraged Nils Åberg[38] to state that here there is 'incontrovertible evidence that Irish millefiori had for the first time been received in Anglo-Saxon art'. There are, of course, earlier examples amongst the prints and escutcheons of Group A bowls; but what the statement emphasises is the Irish origin of the millefiori.

Anything that will show whence inspiration came must not be lightly passed over. We have had glimpses of Irish inspiration in the palmettes of the Barlaston print, and in other minor instances too numerous to mention. But now we come to an instance of a direct connection between the Sutton Hoo escutcheon and the Ballinderry brooch,[39] in that the parallel hatching, the criss-cross and herringbone decoration on the enclosing ring can be exactly paralleled on the hoop of this brooch. This parallel caused Stevenson to proclaim an Irish origin for the hanging-bowl.[40] This statement will rock nobody's boat, since Bruce-Mitford is of the same opinion.[41] Corroborative opinions such as these are important, since they make it much easier to make some future statements and it is for these reasons that the Sutton Hoo bowl has been considered first. Another parallel is to be seen between the large central pattern filled with circles of millefiori on the print, Fig. 79, and the Old Castle of Carbury brooch, Fig. 73:1 as well as Fig. 41:2.

Since the Sutton Hoo bowl was in a closed deposit, it cannot be later than 650. But its condition was such that it must have been very old when it was buried. It had been repaired by the Saxon 'master goldsmith' with pride and care, using precious metals. This is good evidence of the value put upon it, and it goes a long way towards explaining why so many hanging-bowls have been found in Saxon graves. So here we have a hanging-bowl of Irish manufacture, as everyone agrees, treasured by a Saxon king, and buried with other treasures in his boat grave near the coast of East Anglia. And that knowledge is going to make it very much easier to say that the Kingston bowl is also of Irish workmanship.

The Kingston, Kent,[42] hanging-bowl is less hemispherical than are most bowls, and it also has a most pronounced bent-over rim. It was found in a cemetery of 308 graves, in grave 76. The decoration on the small escutcheon (Fig. 81:1) is of a very simple basic character, consisting of three triskeles, all three having arms conjoined at the centre of the disc, thereby giving the impression that a fourth triskele exists at the middle. Patterns such as this are early; and they are seen as the foundation patterns of all others involving the use of four spirals. The idea for this combination of triskeles came from such items as the Trawsfynydd tankard handle,[43] with its swirling triskeles, which Corcoran dates to the latter half of the first century AD. Later, there are triskeles in combination on the open-work sword mount (Fig. 41:1) from Lambay Island,[44] and dated by associated dolphin brooches to the second century. Enamelled equivalents of the Kingston disc occur at Dunning, Perthshire[45] and at King Mahon's fort, near Ardagh, Co. Roscommon,[46] the last being associated with a zoomorphic penannular brooch.[47] King Mahon's fort was a minor heroic centre, and as such its history probably ended at about the same time as did those of the major centres, namely about the middle of the fifth century. The brooch came from a very thin sealed occupation level, and from this same layer came an ibex-headed pin.

Ibex-pins are notoriously unreliable for dating purposes. Reginald Smith[48] dated them to the first century AD, a date supported by one from Dunfanaghy, Co. Donegal,[49] where it was found in association with a first-century Roman brooch, which could have been an heirloom, and therefore is unreliable for dating purposes. Stevenson[50] suggests (rightly) that the type is Scottish, and claims a fourth-century date for it. The zoomorphic pen-annular brooch is an early form, and is most likely to date from the late third century. Everything points to the Kingston bowl being of fourth-century date, which is also in entire agreement with Kendrick's date for it,[51] which he based on the fact that the bowl was old and patched and worn when it was buried.

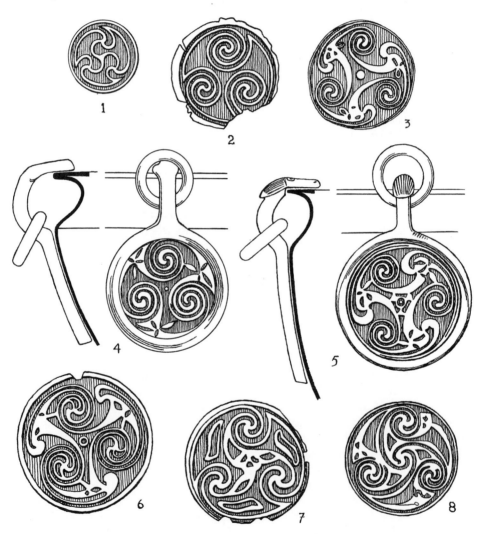

Fig. 81: Bronze Hanging-bowl Escutcheons and Prints 1. Kingston, Kent. 2. 7. Stoke Golding, Leicestershire. 3. Barrington, Cambridgeshire. 4. Lowbury Hill, Berkshire. 5. Winchester, Hampshire. 6. Oving, Buckinghamshire. 8. Camerton, Somerset. (3/4)

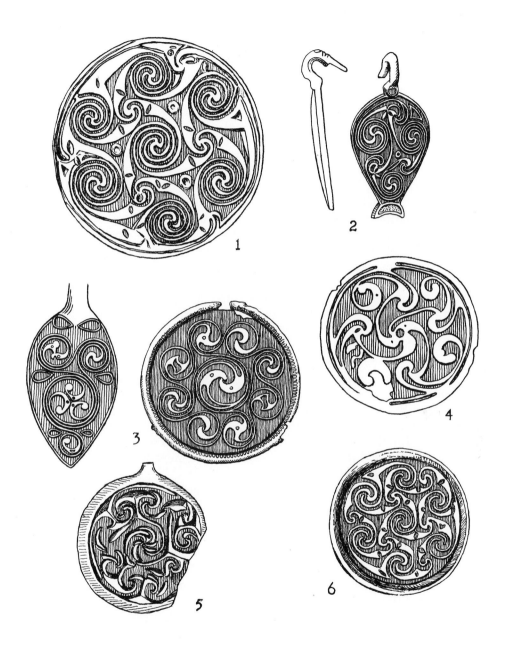

Fig. 82: Bronze Hanging-bowl Escutcheons and Prints 1. Willoughton, Lincolnshire. 2. Benniworth, Lincolnshire. 3. Faversham, Kent. 4. Barrington (Edix Hill Hole), Cambridgeshire. 5. Kaupang, Vestfold, Norway. 6. Winchester, Hampshire. (3/4)

A measure of agreement is beginning to emerge amongst those who have given thought to the problems that these bowls pose. One can see (as if one did not know it before) whence the makers of the Group A bowls were getting their ideas. The Group A attitude was in strong contrast to the consistency of the Group B craftsmen, who, by way of contrast, had behind them a long artistic tradition, and were thus able to draw from their stock of patterns carefully stored over the centuries. The assiduity which they gave to these patterns rather suggests they were members of a guild, jealously guarding their trade secrets, but selling their services to whoever was willing to pay for them. There was a metalworker's shop at King Mahon's fort, so that at least we know where discs like those on the Kingston bowl were being made. But this does not imply that the bowl was made here, for everything depends on the current usage of a pattern, and, if popular, it could turn up anywhere. What we mean here is that we know of at least one site where this particular pattern was being used to decorate bronze objects; and that it was current about the end of the third or the beginning of the fourth century. It will be realised that information such as this is very pertinent to the history of the hanging-bowl in Britain, for it looks to have been a case of the Irish craftsmen working and selling their products in that country. Mlle Henry felt that these men may have been working in ecclesiastical workshops, which is quite feasible, since one such workshop was found at Whitby, where hanging-bowls and skillets were both Irish. But, of course, this is later. However, it must be remembered that Christianity had become the religion of officials in Britain, and the early fifth century saw it making inroads in Ireland; but the British foundations were earlier, and this may have caused a drift of workers to Britain.

These are thoughts in passing, but they are put down in the hope of a better understanding as to why Fig. 81:3 and 4 should be similar to Fig. 76:3 and 5. Fig. 81:3 comes from Barrington, Cambridgeshire, whilst Fig. 81:4 comes from Lowbury Hill, Berkshire,[52] where it accompanied a warrior who was buried with a rather late form of Saxon shield boss. The Oving, Buckinghamshire,[53] escutcheon's spiral-and-scroll arrangement (Fig. 81:6) is precisely that of the Lagore disc. The setting out of the Irish spirals mostly omits the central motif, for the simple reason that it is based directly on the triskele, though there are exceptions (see Fig. 70:3). The same arrangement appears on the bird-shaped escutcheon from Benniworth, Lincolnshire[54] (Fig. 82:2).

One gets taken up by these escutcheons because one appreciates their decoration, and the skill necessary to achieve these well regulated designs; and then one suddenly remembers that, in the whole course of the entire series not one concession has been made to Christianity. If the bowls were put to the uses claimed for them, is it not strange that at least one little cross was not engraved on at least one little escutcheon, as might have been expected. Christian symbols appeared on escutcheons of the Group A bowls, and they appeared on one or two zoomorphic penannular brooches of Irish origin; but even the hanging-bowls discovered at Whitby showed no such concessions: yet one print here (Fig. 78:7) includes a hippocampic head, an Irish trademark if ever there was one.

No typological series is evident in any arrangement of prints and escutcheons, and none is intended, since these objects neither date themselves nor are their associations of any value. But, remembering the very Irish layout of the Oving escutcheon's decoration, it should be compared now with the many spiralled print from Willoughton[55] (Fig. 82:1). Look carefully, and you will see that the whole pattern is composed of three of these Oving layouts, the central triple spiral being common to all three. The craftsman intended us to know this, because he was thoughtful enough to include three little round hollow islands of metal in the enamel background, to remind himself and us of how this large pattern was made up. Further elaboration followed, as seen in the Winchester[56] print (Fig. 82:6). At first glance this looks like an improvement in workmanship; but look closely, and it will be seen that the spirals are breaking up. This print must be studied along with the escutcheon of Fig. 81:5, from the same bowl. The pattern here differs a little from that at Oving, for the spirals are double, like the Lowbury and Barrington discs. Clearly, there is a strong family relationship here, and the watchword was that one design sufficed for every occasion.

In this account there is the makings of a continuing story; but its finale is not written in Britain but in Scandinavia, where, at Kaupang, Vestfold, Norway,[57] the largest collection of insular bronzes ever found in the north has been found in Viking cemeteries. David Wilson[58] has placed the animal style present in 'an Irish Sea cultural zone — almost certainly in a Northumbrian workshop'. The most monumental piece found here was a bronze bowl with a diameter of 32.4 cm. The bottom is convex, and it has a turned-in rim with two pinched-up circles below it. The bowl was found in a boat grave. Three ornamented objects came from a neighbouring grave, one of which was the escutcheon shown in Fig. 82:5. As will be seen, this escutcheon bears somebody's excuse for curvilinear patterns, and there is no trace of enamel. From the grave that produced this escutcheon came a 'pure Anglo-Saxon Northumbrian piece of work'.[59] But the escutcheon is not Northumbrian, nor is it of Irish manufacture; but it must have been made locally in Scandinavia. At Kaupang, refuse ingots, bars, crucibles, moulds, tools, all testify to the former existence of a workshop, and this is where this escutcheon must have been made. The basis of the pattern has clearly been something like the Winchester design (Fig. 82:6), but what a real mess has been made of the pattern! On the right-hand side (Fig. 82:5) there is a fair representation of a double spiral; but look again at what is meant to be a triple spiral at the centre. Who was he fooling, this fellow, but the locals knew of no better. David Wilson was right when he said: 'why should there not have been a local bronze bowl industry?' Clearly, the Irish guiding hand was missing.

David Wilson's reference to a Northumbrian workshop brings to mind two escutcheons which are virtually identical. These are the specimens from Chalton, Hampshire[60] and from Greenwich[61] (Fig. 83:1 and 2). In each case there is only one triple spiral, and this is at the centre of the disc. The other spirals are singles. It is most unusual to find single spirals on escutcheons, and those on the Chalton escutcheons have swollen finials, in the Irish manner (compare with Fig. 67:1b). But now there seems to be another influence at work, in addition to the direct Irish one, and this concerns the ornamental 'D' that occurs three times in the patterns on both the Chalton and Greenwich escutcheons. The only parallel that can be found for this odd representation is amongst the triskele decoration on the Kirkby Thore disc-brooch (Fig. 4:4) which is in the Aesica style. Here the 'Ds' are floating around unattached: attached, they can be seen on the Richborough brooch (Fig. 4:7). However, these brooches are too early for their decorative quirks to have affected the patterns on these two escutcheons; so the source of the idea remains a mystery. All that is positive is the source of both specimens, since they must have originated in the same workshop.

An account such as this is bound to be somewhat disjointed, since there is no real connecting story. All one can offer is a series of vignettes — a collection of near truths and suppositions, but including snippets of positive information, such as the fact that the two escutcheons from Camerton, Somerset[62] (Fig. 81:8 and Fig. 83:3) were found in a

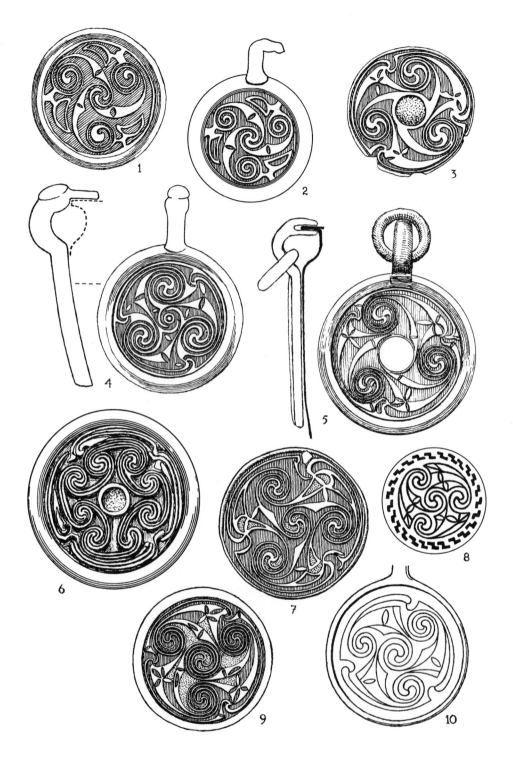

Fig. 83: Bronze Hanging-bowl Escutcheons and Prints 1. Chalton, Hampshire. 2. Greenwich, Kent. 3. Camerton, Somerset. 4. and 6. Chesterton-on-Fossway, Warwickshire. 5. Loveden Hill, Lincolnshire. 7. near Oxford. 8. Lullingstone, Kent. 9. Hitchin, Hertfordshire. 10. Middleton Moor, Derbyshire. (3/4)

girl's grave. So, they had been worn as trinkets. But uncertainty occurs when we ask ourselves: did both discs belong to the same bowl? And we must suppose that they did not. The reason for this supposition is that the workmanship of the first falls short of that of the second disc, which by comparison is more sophisticated, with the triskele arms bearing to the left in contradistinction to the first, on which the arms bear to the right. The conjoined trumpet terminations in the second disc are different from those in the first, and also a third leaflet has been added in the angle formed by the junction, and at right angles to the trumpet ends; whilst a fourth has been inserted into the red enamel filling. The introduction of leaflets assists in the isolation of each of the spirals. So, on balance, two bowls must be represented here, purloined by the parents of the girl: but where did they steal them? Camerton is a long way to the west for Saxons, and we have no means of finding out just how far the pioneers' bandwagon had rolled, or whence it had come. So what we have now got is a bagful of information that is of very little use to us, for it tells us nothing about the objects themselves, which was to have been expected when they were carried about by a little girl belonging to a family of land-hungry invaders marching in search of a place in which to settle. All that can be added is that the Stoke Golding[63] escutcheon (Fig. 81:7) has decoration whose style is remarkably like that of the first Camerton disc, and again like it with arms bearing to the right. Stoke Golding is in Leicestershire, which is many miles from Camerton.

The above remarks highlight the veritable uselessness of the bowls' distribution, for it tells us nothing except that the majority of finds have come from east of the Fosse Way. What the distribution does is to tell us about the Saxons, but not about the bowls which they stole. We do not know how mobile they were; but it is general for people endeavouring to settle in a new country to have to move about a bit, and, in addition, Saxon habits did not help very much either. All we know is that they formed a deep regard for the bowls which they stole, and they treasured them sufficiently to express a wish to be buried with them.

However, some of the best pieces are yet to come. Two enamels from Faversham, Kent,[64] are, without doubt, the work of an Irish hand (Fig. 82:3). Both the escutcheon, which is flat backed, and the print are decorated to the highest standards. There was a preoccupation here with both zoomorphic and ornithomorphic motifs. The decoration consists almost entirely of closely coiled spirals, which take several forms. The differences concern their finials: where these are thickened, with the help of a dot they become ornithomorphic finials, particularly in the case of the escutcheon, upon which they make up a triple spiral, with their heads back to back; a double spiral, with the heads beak to beak; and a lone single spiral. There is also a true double spiral which does not simulate anything else. All this is upon a heater-shaped escutcheon which, according to the Traprain Law evidence, suggests an early period for this piece. The ornithomorphic finials have parallels on the Bloomfield brooch, from Co. Roscommon[65] (Fig. 72:3); on the latchet, Fig. 67:2, and the handpin, Fig. 72:4. Rather less well-defined heads are on the proto-handpins of Fig. 68, all of Irish manufacture. It will be recollected that one of these — the Oldcroft pin — had a *terminus ante quem* dating of AD 354–9, and that it was reckoned to be over 100 years old when buried. On this score, one could place the ornithomorphic style in the mid third century; but the form of head with the little circle at the beak-tip must be later, and for this a fourth-century date would seem to be appropriate.

There are no ornithomorphic finials on the Faversham print except for the centre two, which, beak to beak, form a double spiral here. Note how the necks are hollowed. Elsewhere, amongst the running spirals which make up the rest of the decoration there are two hippocampic heads, which by having necks, snouts and ears extended are made to form triple spirals. Elsewhere in the chains there are triple spirals and double spirals which conform to the normal pattern. In Ireland these hippocampic heads amount almost to a trade mark among Irish products. Copied from the early-second-century Brigantian form, they first appeared on the Petrie Crown and the Bann disc; then they were seen on the latchet, Fig. 67:2 as well as handpins, Fig. 67:3 and 4, and persisted after the seventh century as may be seen from Fig. 76:6. Clearly this form of head is useless for dating purposes, but it does prove Irish provenance. The style of the Faversham print and escutcheon is such that it is just possible that this bowl came from the same workshop as did the Sutton Hoo bowl.

A combination of excellent draughtsmanship and first-class workmanship combine to make the Hitchin, Hertfordshire, disc one of the best (Fig. 83:9). Here we are looking, perhaps for the last time, at what Kendrick called the billowing, sinuous eccentricity of an early masterpiece. The drawing here is done with such ingenuity and delicacy that one can imagine it to have been done on paper. And perhaps this is the reason why this piece has had its pattern compared with the scrollwork in the Book of Durrow.[66] The Book of Durrow dates from about 650. Leeds[67] did his best to prove that this book was the work of an English school, a contention since supported by Lowe on palaeographic grounds, and by Bruce-Mitford on the decoration. The years 635–64 have been specified as the beginning of the Irish Missions to the Saxons. But this is too late a period for the Hitchin disc by as much as one century. We have come a long way in order to gaze at this fine piece, in its red and yellow enamel, which even the Oxford escutcheon cannot touch (Fig. 83:7). Both these pieces are the most advanced of what Kendrick called the developed trumpet-pattern series. From henceforth, the basic form of the triskele is lost by the break-up of the central spiral, though this had not as yet occurred when the Middleton Moor[68] escutcheon was made (Fig. 83:10) for it is most likely to be of the same period as the Winchester bowl, an opinion based on the similarity between the decoration on the print of the latter and an enamelled ornament[69] which was found in the same Derbyshire tumulus. But the break-up had occurred when the second Camerton escutcheon was made (Fig. 83:3), and the process was carried a little farther at Lovenden Hill (Fig. 83:5) on which escutcheon there is now a void at the centre. Haseloff[70] concluded that this feature of the central circular void represented a later stage in development; though to K.R. Fennell[71] this is not true. Haseloff is possibly nearly right, though his suggestion would put such an escutcheon late in the series. But we are still within the pagan Saxon period, before proselytising activities began with the arrival of Fursey in East Anglia in the year 635.

It is at this point that a curious change takes place. We can see its consequences in the print from Whitby[72] (Fig. 78:7) on which there is a circle of badly executed running spirals. This print is of chronological significance, since there is a *terminus post quem* date of 657 for it, which is the year of the foundation of the monastery. So we are dealing with something that could have been made just a generation after Fursey's arrival in East Anglia. Another print (Fig. 78:6) from Bekesbourne, Kent[73] is remarkably like the Whitby specimen, but the workmanship is vastly superior. The same thoughts persisted at Bekesbourne and at Whitby, but both were feeling the influence of the developed trumpet-form pattern. The Irishness of all this is confirmed by the appearance, at the centre of one of the Whitby spirals, of a hippocampic head. The situation that had probably developed by this time was one that witnessed the virtual disappearance of the triskele-based spiral design, so far as metal is concerned; or should one qualify that statement by saying 'so far as hanging-bowls are concerned', because there are two very satisfactory representations of triskele-based spirals on the reverse side of the Tara brooch. Also, a design similar to that on the Hitchin disc was neatly translated into a calligraphic representation in the Book of Durrow. What had happened here? The answer is that craftsmen, who had proved themselves to be excellent draughtsmen on metal, now downed their metalworking tools to take up pen and brush. So the situation appears to be that the apex in hanging-bowl making was achieved at about the end of the sixth century, or at the latest at the beginning of the seventh century: and that this situation, marked by waning activity, was brought about by increased missionary activity in Anglo-Saxon England in the years between 635 and 664. The demand for manuscripts probably outstripped the demand for hanging-bowls, or maybe customs were changing. Activity extended to education, in which Ireland occupied a paramount position in the seventh century.

A clearer picture now emerges. During the latter part of the Occupation, it appears that the Christian Church was fairly well established in some urban areas, in which the new faith became the religion of officials. However, it seems there was another movement which involved the people, particularly those living outside the urban areas; and, according to Charles Thomas, the permissible label for these congregations of the people is 'the Church in Celtic-speaking Britain'. The Irish church, with which it is sometimes confused, did not exist at this time. Although there would appear to be a distinction between the two churches, both were thoroughly Roman in both creed and origin: they differed only because the congregations of the first were made up of Romano-Britons and officials, whereas in the second the congregations were entirely Celtic. For the first-mentioned Church, Romano-Britons made the hanging-bowls of Group A; and for the Church in Celtic-speaking areas of Britain Celtic craftsmen made the hanging-bowls of Group B. Romano-Britons favoured Christian symbolism, whereas among the hanging-bowls of Group B not a single Christian symbol can be found, though Leslie Alcock has suggested that the triskele might in itself have acquired added significance as a symbol of the Trinity.

Comparisons serve to indicate a widening gap between the urban churches and the churches in Celtic-speaking parts of Britain, and this situation is reflected in the metalwork. It was the Church of the urban areas which suffered most after the Withdrawal, whereas the priests of the churches in the remote Celtic-speaking areas continued with their proselytising activities. Some of these priests were bishops, like St Ninian, who had his see in the Isle of Whithorn, where he founded the missionary centre of Candida Casa *c*. 400; and from here a trail of Ninianic foundations extended to different areas, even to as far off as the north of Scotland,[74] and to Nendrum, Co. Down, Ireland (founded 445). Since the Celts were truly individualists, some of these churches were known by the names of their founders (like the Ninianic Church), even though they remained Roman in both creed and origin.

Missionary activity promotes a lot of coming and going, with the establishment of new congregations. With the declining importance of the heroic centres, metalworkers, in their search for new business, must have realised that in these congregations there were opportunities for the practice of their craft. In fact, the advent of Christianity was entirely opportune, in that ecclesiastical centres provided not only shelter, but security, with the added opportunity, as at Whitby, of founding workshops. As Wheeler[75] quite rightly pointed out many years ago, the Celtic temperament was a plant of tender growth, flourishing only in moments of security. The Church provided that security, with the added opportunity for the manufacture of votive chalices. Undoubtedly, the spread of Christianity to many parts of Britain helped in the Revival as we know it, so that once again superb works of art were freely available. The scarcity of early examples of hanging-bowls can be ascribed to the turmoil that resulted from the Withdrawal and the inroads of the early Saxons.

However, around 500 the Britons divested themselves of the Saxon yoke, at the battle of Mount Badon, and thereafter there was peace for two generations until 550, when the Saxons returned. By 600 most of England had fallen to the Saxon kingdoms. So, for about half a century there was peaceful coexistence, and these conditions permitted the art to flower again. To this period must be assigned the bulk of the hanging-bowls. That we have these bowls at all was due to the sacking of religious foundations and the acquisitive nature of the Saxons. Events have shown that the Irish were better placed to make peace with the Saxons; and when the Saxons themselves were converted to Christianity, Irish craftsmen, who had followed in the wake of their missionaries, found that political and economic security necessary for the exercise of their talents. This would appear to be the story which the hanging-bowls have to tell.

NOTES

1. K.R. Fennell, 'Hanging-Bowls with pierced Escutcheons', *Med. Arch.*, IV (1960), p. 127.

2. F. Henry, 'Hanging-Bowls', *Journ. Roy. Soc. Antiq. Ireland*, LXVI (1936), pp. 209 ff.

3. Peers and Ralegh Radford, 'The Saxon Monastery of Whitby', pp. 27 ff.

4. D. McRoberts, 'The Ecclesiastical Significance of the St. Ninian's Isle Treasure', *Proc. Soc. Antiq. Scot.*, XCIV (1960–1), p. 304.

5. At monastic foundations dedicated to St Nektarios.

6. R.A. Smith, 'Two Early British Bronze Bowls', *Antiq. Journ.*, VI (1926), p. 277. Figs. 1–2.

7. T.D. Kendrick, 'British Hanging-Bowls', *Antiquity*, VI (1932), pp. 161 ff.

8. National Museum of Antiquities of Scotland.

9. A.O. Curle, *The Treasure of Traprain* (Glasgow, 1923), 78, pl. XXX.

10. Henry[2], pl. XX:3.

11. Ibid., pl. XX:1.

12. Ibid., pl. XX:2.

13. Kendrick[7], Fig. 2.

14. In an early-third-century deposit. R.E.M. and T.V. Wheeler, 'The Roman Amphitheatre at Caerleon, Monmouth', *Arch.*, 78 (1928), p. 164, Fig. 14:17.

15. Bushe-Fox, 'Third Richborough Report', p. 78, pl. IX:13.

16. Attributed to AD 150–250. *Arch. Ael.*[3], VII (1911), p. 184, Figs. 20–1.

17. Ibid., 184.

18. Henry[2], p. 218.

19. Leeds, *Celtic Ornament*, pp. 154 ff.

20. J. Romilly Allen, 'Metal Bowls of the late Celtic and Anglo-Saxon Periods', *Arch.*, LVI (1898), p. 40, Fig. 1.

21. A. Small *et al.*, 'Craig Phadraig', *University of Dundee Dept of Geography Occasional Papers*, I.

22. *Med. Arch.*, VIII (1964), p. 236, pl. 19:C and D.

23. H.E. Kilbride-Jones, 'A Bronze Hanging-Bowl from Castle Tioram, Moidart; and a suggested absolute Chronology for British Hanging-Bowls', *Proc. Soc. Antiq. Scot.*, LXXI (1936–7), p. 207. Fig. 1 and Fig. 2:1.

24. T.C. Lethbridge, 'Bronze Bowl of the Dark Ages from Hildersham, Cambs.', *Proc. Cambs. Antiq. Soc.*, XLV (1952), p. 44, Pls. IX and X.

25. E.T. Leeds, 'An Enamelled Bowl from Baginton, Warwickshire', *Antiq. Journ.*, XV (1935), p. 1.

26. Kilbride-Jones[23], Fig. 2:3.

27. Bushe-Fox, 'Second Richborough Report', p. 33, pl. XIX:33.

28. Romilly Allen[20], p. 43, Fig. 4.

29. E. Fowler, 'Celtic Metalwork of the Fifth and Sixth Centuries A.D.', *Arch. Journ.*, CXX (1963), pp. 98–160.

30. Romilly Allen[20], p. 43, Fig. 4.

31. Leeds[19], p. 147.

32. Kendrick[7], p. 173.

33. Romilly Allen[20], p. 48, Fig. 7.

34. T.D. Kendrick, *Antiq. Journ.*, XXI (1941), p. 236.

35. Kendrick[7], Fig. 169, Fig. 5; R.A. Smith, *Proc. Soc. Antiq.*[2], XXII (1907–9), pp. 66 ff.

36. K.R. Fennell, 'Lovenden Hill, Lincolnshire', *Med. Arch.*, I (1957), pp. 148 ff.

37. R. Bruce-Mitford, *The Sutton Hoo Ship Burial* (BM, 1972), pl. C.

38. N. Åberg, *The Occident and the Orient in the Art of the Seventh Century* (KVHAA), Handbook 56: I, p. 56.

39. H.E. Kilbride-Jones, 'Zoomorphic Penannular Brooches', *Soc. Antiq. Research Comm. Report*, XXXIX.

40. R.B.K. Stevenson, 'The Hunterston Brooch, and its Significance', *Med. Arch.*, XVIII (1974), p. 31.

41. Bruce-Mitford[37], p. 86.

42. Faussett, *Inventorium Sepulchrale*, p. 55, pl. XXI:5. Also *Proc. Soc. Antiq.*[2], XXX (1917–18), p. 86, Fig. 28.

43. J.X.W.P. Corcoran, 'Tankards and Tankard Handles of the British Early Iron Age', *Proc. Preh. Soc.*, XVIII (1952), p. 97, pl. XII:1.

44. R.A.S. Macalister, 'On some Antiquities found on Lambay', *Proc. Roy. Irish Acad.*, XXXVIII C (1928–9), p. 243.

45. Macgregor, *Early Celtic Art in North Britain*, no. 260.

46. Unpublished. Information by courtesy of L. de Paor.

47. Kilbride-Jones[39], no. 60.

48. R.A. Smith in *Proc. Soc. Antiq.*[2], XX (1903–5), p. 350.

49. T.B. Graham and E.M. Jope, 'A Bronze Brooch and Ibex-Headed Pin from the Sandhills at Dunfanaghy, Co. Donegal', *Ulster Journ. Arch.*[3], 13 (1950), p. 54.

50. R.B.K. Stevenson, 'Pins and the Chronology of the Brochs', *Proc. Preh. Soc.*, XXI (1955), p. 291.

51. Kendrick[7], p. 183.

52. D. Atkinson, *The Romano-Site on Lowbury Hill, Berkshire* (University College, Reading, 1916), pp. 18–21, pl. V.

53. *V.C.H. Bucks.*, I, p. 195.

54. T.D. Kendrick, 'A new Escutcheon from a Hanging-Bowl', *Antiq. Journ.*, XVI (1936), p. 98.

55. T.D. Kendrick, 'Escutcheon of a Pagan Saxon Hanging-Bowl from Willoughton, Lincs.', *Antiq. Journ.*, XXV (1945), p. 149.

56. W.J. Andrew and R.A. Smith, 'The Winchester Anglo-Saxon Bowl', *Antiq. Journ.*, XI (1931), p. 1.

57. C. Blindheim, 'A Collection of Celtic Bronze Objects found at Kaupang, Vestfold, Norway' in B. Almqvist and D. Greene (eds.), *Proc. of Seventh Viking Congress* (1976), pp. 9 ff.

58. A. Small, C. Thomas and D. Wilson, *St. Ninian's Isle and its Treasure* (Oxford, 1973), I, pp. 86 and 129.

59. Blindheim[57], p. 14.

60. P.V. Addyman *et al.*, 'Anglo-Saxon Houses at Chalton, Hampshire', *Med. Arch.*, 16 (1972), pp. 13–31.

61. Leeds[19], p. 150, pl. III:5.

62. D.E. Horne, 'Celtic Discs of Enamel', *Antiq. Journ.*, X (1930), p. 53.

63. A.J. Pickering, 'A Hanging-Bowl from Leicestershire', *Antiq. Journ.*, XII (1932), p. 174.

64. Romilly Allen[20], Fig. 7.

65. Kilbride-Jones[39], no. 101.

66. Henry[2], pl. XXXII:9, illustrates a detail from the Book of Durrow which compares fairly well with the Hitchin decoration.

67. Leeds[19], p. 158.

68. S. Pegge, 'Discoveries in opening a Tumulus in Derbyshire', *Arch.*, IX (1789), p. 189, pl. IX.

69. Kendrick[7], pl. VIII:2.

70. G. Haseloff, 'Fragment of a Hanging-Bowl from Bekesbourne, Kent, and some ornamental Problems', *Med. Arch.*, II (1958), pp. 72 ff.

71. Fennell[1], p. 127.

72. Peers and Ralegh Radford[3], pl. XXVIII:5.

73. Haseloff[70], p. 72.

74. There is evidence of escutcheon production in Orkney, in the occurrence at the Broch of Birsay, in a pre-Norse level, of a lead disc suitable for casting bronze escutcheons. There is ample evidence of metalworking: clay crucibles, clay and stone moulds, bronze bars and pieces of coloured glass. C. Curle, 'An engraved Lead Disc from the Broch of Birsay, Orkney', *Proc. Soc. Antiq. Scot.*, 105 (1972–4), p. 301, Fig. 1 and pl. 24.

75. R.E.M. Wheeler, 'The Paradox of Celtic Art', *Antiquity*, VI (1932), p. 298.

27. Epilogue

The culture of a civilised people is represented chiefly by works of art. The objects examined in the foregoing pages are representative of the intellectual development of a linguistic/political/ethnic group of people called the Celts: archaic objects decorated with eccentric curves and swelling forms, an art of abstract designs, practically all made out of a single line; and an art that is at once distinguishable from all it rivals. Some of the individual pieces are so perfect, so entirely rich and harmonious in design, that those who made these designs must be considered to have been amongst the foremost artists of this world. Although 'Celtic' art has no roots of its own, it is the art of individuals, and is so individual in itself that it is something of a paradox, in that it was probably the main cultural force that bound this linguistic/political/ethnic group together. It is not a personal art, in the sense that it does not run to statues in the village squares, nor does it embrace more than a few small representations of the human face. But, as Wheeler remarked many years ago, to attempt to explain the genesis of something like Celtic art is like trying to explain the genesis of the bumble-bee. It can be explained only as the art of a conglomerate of peoples whose domains stretched almost from the Black Sea to the Atlantic and included Britain. After it had crossed the Channel to Britain it developed an individuality of its own that might have seemed strange to a continental; though, of course, its source remained continental.

'Celtic art' is the readily understandable term for this art, which has been used throughout this book. This art was slow in development until new life was injected into it by the arrival in the first century BC of Belgic settlers who, apart from developing the rich Hertfordshire lands which the Britons had ignored, also set about broadening the aspect of the art, by introducing new ideas and new techniques of metalworking. So that one might say that native British art and native British culture reached their climaxes about the turn of the century BC/AD. A measure of security was provided by increased wealth, and there was an increase in trade with the Continent. The rising graph of affluence is reflected in luxury articles such as decorated mirrors, the best probably belonging to Romanising Belgic landlords. The Belgae were better acquainted with Mediterranean classical extravaganzas than were the Britons. The magnificently decorated accoutrements reflect this increasing wealth of the British princes, a fact not missed by the Roman spies who came to Britain dressed as merchants. The reports they sent home must have been good, and that is why the Romans came tumbling into the country in AD 43.

Yet, once the Occupation was an established fact, nothing seems to have worked for the common good. The corn crops were requisitioned, and a rapid and disastrous change came over the whole spirit of British craftsmanship. Before long, the tide of Roman mass-production burst upon the Britons. The imperial standard of classical art was a degraded commercialised variety which dominated the minds of those who set the fashion. It must be said, however, that the ancient world was not aware of any distinction between art and manufacture, so the modern conception did not apply. The native craftsman endeavoured to keep going by the quick production of brooches and pins; so that one might say that they exchanged craftsmanship for trade. Eventually, they succumbed to using Roman mass-production techniques themselves, and in the second century they were confronted with a trade boom. Then came the *coup de grace*, the stab in the back, which came, not from the Romans, but from their own people, the barbarians from the north. Economically, the metalworkers were ruined.

Ireland, too, had similar art which also reached its zenith at about the same time as that in Britain. But the Irish appear to have taken a jaundiced view of the Occupation of Britain; and good strong zoomorphic penannular brooches were made for those adventurous spirits who set out to raid its coastal areas. Other articles are scarce in Ireland, and most may have been melted down as a source of raw material, as was being done by the brooch-maker at Clogher. During the whole heroic period from the second to the fifth century, the art was in need of a stimulant. With a narrowing of talent in Ireland, and its disappearance altogether in Britain, there must be a lesson to be learnt from all this. The lesson is that, without national security, the Celtic temperament tended to wither at the roots. From the first half of the fifth century there was ushered in a period formerly known as 'the Dark Ages', in which it was assumed everything came to a full stop. Only 40 years ago Wheeler could write of a 'hiatus' of three centuries in Celtic art in Britain; and it was belief in such a hiatus that gave rise to the hypothetical 'dark ages', and these ages were dark because the situation at that time was not fully understood, due partly to a misdating of some objects and the absence of others since noted.

Also, nobody bothered to study the tactical position of the Christian church in Britain. One would assume that its effect on the art would be far-reaching. Christian graves may be sterile, but the Saxons were conservators of Celtic art; but it was by disaffranchising them of that art that we began to learn more about those who were judged responsible. Now we see a Celtic revival, a gradual recrudescence rather than a sudden breaking out. The impetus must have come from somewhere, and in the general run-down caused by the economic depression in both Britain and Ireland in the early fifth century, the one stable body was the church. The church provided cover and security, and both came just in time to save the art from becoming moribund. Of course, the church itself suffered vicissitudes, but it had the ability to rebound. Whatever the conditions were, religious observance requires the use of plate, chalices and other furniture.

Present architectural remains are no guide to the extent to which Christianity penetrated society. By 312 there were bishoprics at London, York and Colchester; and Cirencester has produced evidence of Christianity in the same period. The situation remained static for most of the first half of the fourth century, but then came the revival of the old Celtic religion. The great pagan temple at Lydney was built some time after 364, with some official help. However, the church was gaining ground among the villa-owning aristocracy, and in the cantonal towns, for the armies and their leaders were at this time Christian. By the beginning of the fifth century the Church in Britain was producing well educated men who were capable of making their mark in the wider world of western Christianity. All the evidence points to the fact that this Church made rapid but not spectacular growth during the first half of the fifth century. There were two zones of activity: one was the urban central region of the old Civil Province; the other was the frontier zone, which took in Galloway.

However, the church in Celtic-speaking Britain set out to dominate the north. Apparently, the tribes had little sympathy for the episcopally organised religion of the south. Formerly, great emphasis was placed on the proselytising activities of Ninian, but now his contribution tends to be minimised. It is known, of course, that there were distinct traces of Christianity in the Carlisle area in the fourth century, so that one cannot deny a spiritual leader to the people who lived here. But, by the early fifth century, Carlisle was within the kingdom of Rheged, with Christian communities living on both shores of the Solway. Clearly, there must have been a Christian community at Whithorn, and since St Ninian is known to have been a bishop he was probably sent to be over the Christians resident in this area. Bede's statement that Ninian was a bishop in fact implies that there was a Christian community in Galloway before Ninian was consecrated. Whithorn was well within the territory of the Novantae, within which there was also an immigrant Irish population; and the connection with Ireland went further when Nendrum, in Co. Down, was founded in 445 by St Mochaoi. This was a significant moment in Irish history, since the heroic centres were fast falling apart at this time.

It has been said that traders and artisans played their part in the spread of Christianity. Their importance was in their ability to copy and to disseminate Christian symbols, which would excite the curiosity of those seeing them for the first time. So far as the Irish craftsmen were concerned, the arrival of the church must have been in the nick of time, since it gave them continuing opportunity for metalworking. The church in the north and west was a diocesan church, and presumably each diocese must have enjoyed a certain amount of autonomy in these times. Church plate and votive offerings must have kept the metalworkers busy, and the prospects overseas must have emptied Ireland of her craftsmen. This is the only possible explanation for the slow, but nevertheless dramatic revival of Celtic art in Britain: it had been a gradual process beginning some time during the later fourth century. It is possible to imagine the 'groaning' Britons, after the receipt of the rescript of Honorius, turning to the only stable body left, the church, for a measure of solace and stability. This is possibly the explanation for the rapid growth of the church in Britain in the first half of the fifth century; and it was on this wave of expansion that St Machaoi arrived at Nendrum. The British church in the south could even afford the luxury of a heresy; whilst, on the other hand, there were well organised pilgrimages to the shrine of St Alban, such as that made by Germanus of Auxerre in 429.

The one representative article in the renaissance in Celtic art in Britain is the hanging-bowl. If, as it seems possible, these were votive offerings given to the church in fulfilment of a vow, then their donors must have been men of means. We have seen how the bowls themselves divide naturally into two types. Clearly one group was made for the church in urban areas, whilst the other, with notably Celtic characteristics, was made for the church in Celtic-speaking Britain. Strangely, the one country which has produced manufacturing evidence is Caledonia, or Pictland, as some would have it, generally considered to have been outside the main areas of Christian influence, though there are Ninianic foundations in the north. A clay mould for casting an escutcheon was found not far from Inverness, whilst another mould, this time of lead, was found in Orkney.

In the wars with the Saxons, although there was a period of Saxon success from 440 to 457, from about 460 British resistance to the Saxon inroads was on the increase, and there followed a period of British dominance from about 495 until 570. But, in 570 the position was reversed, and by 600 virtually all of the former Civil Province was under

Saxon control. This final Saxon success probably included the rifling of the churches.

An outburst of missionary expansion broke out from Ireland in the years between (say) 550 and 650. That this was possible indicates that the Irish had good relations with the Saxons. It was in the year 635 that the Irish missionary Fursey went to East Anglia, where he was received with honour by the king, and where he founded a monastery in Burgh Castle, near Yarmouth. Lindisfarne was founded in the same year, and Melrose shortly afterwards. Hartlepool was founded in 640 and Whitby in 657. If the metalworkers followed hard on the heels of these missionaries, one can imagine an outburst of metalworking at this time.

In any case, there must have been already a long run of very competent design work on metal by the mid seventh century, for so sophisticated a production as the Book of Durrow (*c.* 650) to have been illuminated with designs, some of which are akin to those on the later escutcheons. If the artists abandoned metal for the calligrapher's art, who can blame them, but this does not constitute a hiatus. Later, with waning Irish influence as a result of Saxon cultural inroads, and with the growth of the church in Ireland, emphasis is once more returned to that country, resulting in the flowering of the art in the eighth century.

This epilogue was written for the purpose of examining the reasons for the revival of Celtic art in the years following upon the Withdrawal. We have seen that the impetus for that revival was already there before 410, because it is now realised that the revival was associated with the spread of Christianity. The church in the urban areas was the first in these islands to get established, when it became the government religion as far back as 311. But, whereas the 'official' church suffered under Saxon repression, the church in the Celtic-speaking areas, having better relations with the Saxons, made inroads into the country from the north-west, bringing with it its craftsmen whose skills are enshrined in the enamelled hanging-bowl escutcheons and prints. These skills were still very much in evidence in Britain even in the seventh century, at which period our story ends.

Index